WESTE[RN] TIMES

BR (W)

ISSUE No.9 - SPRING 2024

CW01652032

Contents

The Transport Treasury

TIMES SERIES

Above: The Great Western Railway network was fortunate in traversing many beautiful areas of England and Wales, one of which was undoubtedly the 'Golden Valley' line between Chalford and Stroud on the Swindon to Gloucester route. Having reached journey's end at Chalford in August 1964, a rather grubby lined green Collett 14xx 0-4-2T No. 1458, waits to cross over onto the down platform for a return trip to Gloucester propelling its Hawksworth designed autotrailer. ***Bernard Mills collection.***

Front Cover: The archetypal depiction of a Great Western branch line terminus is provided by 14xx No. 1447 standing at Wallingford with an impending departure for Cholsey in 1958, the penultimate year of passenger operation. Further photographs of the line (now in the hands of preservationists) feature later in this issue. ***Norman Simmonds (NS 305).***

Rear Cover: This image has initiated conjecture amongst the editorial team, as to the identity of the boiler featured. The photograph is believed to date from a visit to Swindon Works in 1957, and is adjacent to the traverser at the eastern end of 'A' Shop. The boiler is sat upon a trolley used to move such items around the works site, from the Erecting Shop into the nearby Boiler Shop. It is thought that the item is from a small pannier tank, due to the mounting bracket seen on the smokebox side. Reader's thoughts and observations to help prove its origin are welcomed.

ISBN 978-1-913251-69-7

First Published in 2024 by Transport Treasury Publishing Ltd.
16 Highworth Close, High Wycombe, HP13 7PJ.

Compiled and designed in the UK. Printed in Tarxien, Malta by Gutenberg Press Ltd.

INTRODUCTION

Mr T R Perkins, a man whose name is possibly less familiar among younger generations, made his mark before World War 2 by travelling the entire railway system of the United Kingdom then open to public services. This achievement can never be equalled as the total route mileage then was vastly greater than today's Beeching-truncated network. TRP's triumph was all the more remarkable in that it was achieved mainly during his annual holidays. Most of his spare time for the rest of the year must have been devoted to poring over timetables to minimise delays between connections to keep on the move. His feat, which took about 12 years and was completed in 1929, stands at the apogee of the determination and resilience displayed by so many who love railways.

The editorial team's activities extend beyond propagation of *Western Times*, and a recent assignment involved the proof-read of a new work based on experiences that have a flavour of the TRP saga. It concerns a railway geographically far removed from the GWR and within the time frame of the 1980s and 1990s i.e. outside the self-imposed 1977 boundary that governs *WT*. The work subjected to the proofing process stands out as the text was drawn from detailed notes, and the photographs were extracted from a collection of over 30,000 images. All this base material was the subject of the author's own labour i.e. the work of one man. Although never a professional railwayman, for several years he held a fully authorised 'all lines' footplate pass which allowed him aboard locomotives that penetrated freight-only routes and rarely used industrial sidings in addition to the lines used by public services. In that respect, this odyssey traversed a wider spectrum of the system's operations than TRP could have managed many years earlier.

Countless hours spent riding locomotives and in lengthy delays awaiting connections made it inevitable that many friendships should have developed with working railwaymen. Thus extended dialogues 'inside the fence' with expert practitioners rendered authoritative explanation and confirmation of conclusions drawn by this 'amateur' observer. The resultant commentary is thus imbued with enviable authenticity because the information was assembled *in real time*.

This attribute contrasts with the processes adopted in the composition of *Western Times*. The GWR with its associated enterprises is probably the most comprehensively catalogued of Britain's major railways, yet much information is effectively second-hand. Review of topics increasingly remote through the passage of years makes accurate capture of the organisation's texture challenging. Deduction or argued speculation risks assumption of the status of proven fact, perhaps erroneously endorsed by repetition in other publications. Admission of ignorance is a useful weapon in enticing those with superior knowledge to correct or to contribute to the back story. Research is immensely enjoyable but has a voracious appetite for time so any measure that helps share the load is welcome. This is why the 'Guard's Compartment' column is so important.

Recognition of contemporary factors is relevant. *WT* Issue 6 included a commentary on the short working lives of Hawksworth coach stock, particularly the suburban vehicles. The following issue included useful information from a reader which *inter alia* contended that inferior damage resistance justified early replacement in favour of BR Mark 1 vehicles. It would be interesting to know whether differing resilience levels warranted premature withdrawal of far from fully amortised capital assets. In the modern world, much emphasis is placed on safety but how relevant a factor was this in the equation at the close of the 1950s? Also, Beeching's conclusions on poor stock utilisation are directly linked with his 1962 report whereas, for example, the scrapping of Diagram E166 Suburban Composite coaches was well under way by December of that year, before the good Doctor's conclusions could have been implemented.

The reader's comments about compartment accommodation on suburban services were apposite, evoking memories of the daily crawl around the Catford loop in a grimy, dusty Southern 4-SUB during the warm summer of 1966. The success of the pre-war GWR diesel railcar fleet was in part due to the saloon-type accommodation and it was unfortunate that this popular feature failed to translate to post-war steam-hauled suburban services. Was this a case of stocking and operating services in accordance with the dictates of tradition without thought for what the paying passenger might prefer?

--- o O o---

Editor: Andrew Malthouse

Editorial Assistant: Jeremy Clements

To contact the editorial team please email: **WesternTimes@mail.com**

For sales, subscriptions and back issues of *Western Times* please go to: **www.ttpublishing.co.uk**

STROUD'S GOLDEN VALLEY

ALAN POSTLETHWAITE

The London & Bristol Railway Act was passed in 1835. The Stroud line followed in 1836, known as the Cheltenham & Great Western Union Railway. There was opposition to the Stroud line, firstly from the London & Birmingham Railway who wanted to link Cheltenham with their existing line at Tring, then from the Thames & Severn Canal Company and finally from Squire Gordon of Kemble. The last two were each compensated for £7,500.

Engineered by Isambard Kingdom Brunel, the broad gauge line opened from Swindon to Cirencester on 31 May 1841 and from the junction at Kemble through to Gloucester on 12 May 1845. Its western end was difficult to build with Sapperton Tunnel at the summit, just over a mile long, and with many timber viaducts along Stroud's Golden Valley. These viaducts were later rebuilt in stone, encasing the original wooden structure.

Brunel also built substantial stations at Brimscombe, Stroud and Stonehouse. Seven halts were later added, served firstly by railmotors and then by push-pull 'auto' trains. The much plainer Chalford station opened in 1897, having been deemed unsuitable originally for restarting trains up the Sapperton Bank when locomotives were less powerful. Chalford additionally had its own engine shed, which was to be destroyed by fire in January 1916, together with one of the resident railmotors. It was never rebuilt.

In 1903, having conducted trials with an LSWR steam railmotor, the GWR built its first railmotors for use on local services between Chalford and Gloucester. They had flat fronts with low windows, making it difficult for some drivers to see. The later railmotors were bow-fronted with tall windows. Until platforms were built at the halts, wooden steps were deployed from each train.

Above: An Up express in the capable hands of No. 5018 *St Mawes Castle* approaches Standish Junction where the LMS and GWR main lines diverge. This is thought to be summer 1954, when the engine was at Gloucester Horton Road. *George Heiron (1284).*

Above: The first station at the western end of the line was Stonehouse (Burdett Road), seen here in July 1960. When scheduled for closure in 1975, it was saved by vigorous local campaigning. Instead, the station was simplified and modernised, but Brunel's lovely stone building was tragically demolished. ***Norris Forrest (NF 044-13).***

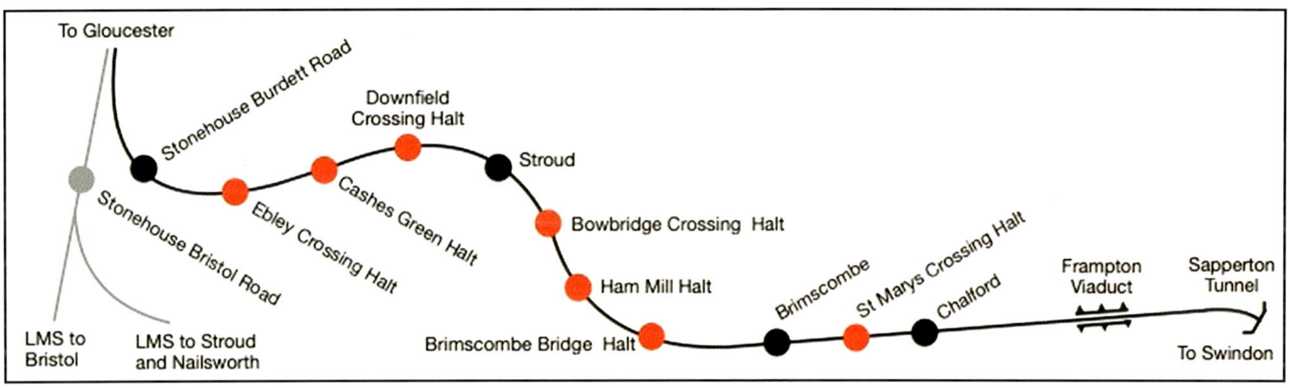

The April 1929 timetable shows a weekday hourly service between Chalford and Stonehouse with about one in three trains running through to Gloucester. The same timetable shows ten Paddington services in each direction. The 'Cheltenham Spa Express' was introduced in 1923 for a fast service to Paddington, challenging the LNER for speed records. The most famous run occurred on 6 June 1932 when the Cheltenham-Paddington afternoon service became the first in the world to be scheduled at over 70 mph.

Under the Beeching axe, all stations and halts closed in 1964 except for Stroud and Stonehouse. Stonehouse has since been rebuilt but Stroud station is preserved, Grade II listed. Stroud goods shed is also preserved but the rest of the goods yard is now a car park.

Stroud's Golden Valley is so-called because of the striking autumnal colours of its beech trees, particularly between Sapperton and Brimscombe. The River Frome runs throughout, once serving multiple mills which were world-famous for their broadcloth. The river is criss-crossed in many places by two railways, two canals and the A419 main road. It is a paradise for industrial archaeologists.

I first photographed the line in 1964 and some of those images are presented here for the first time. Living near the Golden Valley since 1972, I have never tired of exploring the railways, canals, mills, inns, churches, woods and great commons hereabouts. Please enjoy this nostalgic pictorial journey along the line, celebrating a railway set in glorious Cotswold scenery.

Top Left: A train hauled by a 4-4-0, possibly an 'Atbara', arrives at Stonehouse Burdett Road station from the Cheltenham direction in the late 19th or early 20th century. *Lens of Sutton Association.*

Top Right: Steam Railmotor No. 59, which entered service in November 1905, arrives at Ebley Crossing Halt, circa 1910. Opened on 12 October 1903 with simple ground level platforms, these were soon replaced with standard height wooden structures, pagoda shelters and gas lighting. *Lens of Sutton Association.*

Middle: Cashes Green was the final halt to be built on the line, opening on 22 January 1930 to serve a new housing estate. Its rudimentary corrugated metal shelters was much plainer than the usual pagoda style. *Lens of Sutton Association.*

Above: This 1964 image captures a Collett 14xx hauled push-pull train approaching Downfield Crossing Halt, with its grassy platforms and metal pagoda shelters. This halt once served four secondary schools and as a result patronage was buoyant, albeit only at certain times of the day. Downfield was another 12 October 1903 opening, with the introduction of railmotor services. *Author.*

Top: The principle station on the line was at Stroud, and the station forecourt is seen here in the summer of 1964. The frontage constructed in mellow Cotswold stone pervaded a sense of welcome, and the scene is completed by a minibus and shiny Austin A40 Somerset saloon. The I. K. Brunel designed structure is Grade II listed and as such is much unchanged to this day. *Author.*

Bottom Left and Right: Mail is piled high on Stroud's down platform. An LMS designed GUV is being attached to the rear of a Chalford to Gloucester auto train by a 'Large Prairie'. The Collett 14xx Class 0-4-2T would have arrived, propelling the Hawksworth trailer, and as can be seen, departed sandwiched between it and the mail van. This was a regular weekday evening practice on the line. *Author.*

Top Left: Collett 5101 Class 2-6-2T No. 4103 of Bristol Barrow Road shed, storms through Stroud with a mixed freight working from the Swindon direction. By this 1964 date the locomotive was a matter of weeks away from withdrawal. *Author.*

Top Right: From a similar vantage point as the adjacent image, No. 6944 *Fledborough Hall* leads a funeral cortège of three withdrawn locomotives over the double slip from the Stroud goods yard across onto the down line. *Author.*

Above: We view the same entourage again, crossing Merrywalks Viaduct at the opposite end of the station heading west. The trio of unidentified tender engines, consisting of a Castle, 28xx and a Modified Hall, are probably en-route from Swindon Works to one of the South Wales scrapyards, such as Cashmore's of Newport or Woodham's at Barry for final disposal. *Author.*

Top: GWR Steam Railmotor No. 5 arrives at Stroud from Gloucester, circa 1904. This vehicle was from a batch of four (Nos. 3-6) completed at Swindon during April 1904. They provided only third class accommodation, albeit divided into smoking and non-smoking saloons, and were gas lit only for the duration of their service. The coach body of No.5 is recorded as being withdrawn in May 1915, although the engine unit (No. 0805) lasted until September 1919, paired with another body.
Note also the trailing crossover in the centre of the station and the original footbridge.
Lens of Sutton Association.

Middle: Gloucester allocated Collett 14xx Class 0-4-2T No. 1453 stands impatiently 'blowing off' in the Down platform at Stroud. The board has already dropped, so departure for Gloucester must be imminent. These stopping services ran efficiently between Chalford and Gloucester from 1903 until closure of the intermediate stations in 1964, initially with steam railmotors and latterly with loco-hauled autotrailers.

No. 1453 had arrived at Gloucester in July 1962, where it was to remain until withdrawal in November 1964. It latterly lost its smokebox numberplate and was instantly recognisable by the locally fabricated substitute.
Author.

Bottom: After a sprightly dash to the end of the Up platform, the photographer captures the same train crossing Merrywalks Viaduct at the west end of the station. The twelve arch brick viaduct crossing the A46 main road to Bath, was one of the largest civil engineering structures on the line. The siding to the right led to a wagon turntable which served Stratford Mills. There were once plans to build a branch line from here to Painswick, but these aspirations came to nothing.

Note how the down starting signal is located on the opposite side of the running lines, this was to aid sighting for drivers and occurred in several locations in the valley.
Author.

Above: Looking towards Swindon from the end of the Down platform, we see an unidentified 'Western' Class diesel-hydraulic approaching light engine. To the left can be seen the Brunel designed Goods Shed and Stroud Signal Box, a GWR type 27C built in 1905. During the 2010s, the goods shed was restored by the Stroud Preservation Trust to become an arts venue. Beyond the coal siding on the right is the site of the former stables, jointly owned by the GWR and LMS, for carting. The Midland Railway terminus, on the 1¼ mile branch from Dudbridge, was situated fifty yards to the right of picture. *Author.*

Below: The small bay platform at the Up end of the station, plays host to Gloucester Horton Road allocated 'Large Prairie' No. 6128. This 1964 scene is rich with infrastructure detail, an enamel totem affixed to the concrete lamppost, cast iron water column, and the obligatory tubular metal-post bracket starter signal. *Author.*

Right: A series of photographs showing the trio of small halts situated between Stroud and Brimscombe. Opened in conjunction with the introduction of the GWR steam railmotor services, all originally featured the classic pagoda shelters, which as can be seen were later replaced with austere rectangular metal structures. Electric lighting however, was provided from February 1939. All three halts closed on 2 November 1964 following the removal of the Chalford auto services.

Bowbridge Crossing Halt (Top) was opened on 1 May 1905. The distant overbridge carried the minor road of 'Butterow Hill' between Bowbridge and Rodborough Common. This was one of the halts to receive replacement brick and stone platform faces.

Ham Mill Halt (Middle) opened on 12 October 1903 and was originally named Ham Mill Crossing Halt, before the name was shortened as late as July 1957. Like most of the intermediate halts on the line, no trace remains today of its very existence.

Brimscombe Bridge Halt (Bottom) opened on 1 February 1904. Situated less than a mile to the west of Brimscombe village, it featured staggered platforms either side of a stone overbridge that carried the Brimscombe Hill road to nearby Minchinhampton.

Lens of Sutton Association.

Below: No. 1453 is seen again, this time with a pair of Hawksworth designed autotrailers entering Brimscombe on Saturday 24 October 1964. The photographer recorded the train as the 4.40pm departure to Chalford. The rather unkempt nature of the platforms foretell the imminent closure of the station a week later, following the withdrawal of local stopping passenger services on the line. *Peter W Gray (PG 4004).*

Above: Class 9F No. 92244 hauls an Up coal train out of the goods loop at Brimscombe on 24 October 1964, ready for the climb up to Sapperton Tunnel. 'Large Prairie' No. 6113 can be glimpsed through the steam waiting to provide banking assistance. In case a preservation order arrived, demolition of the station took place immediately after closure. *Peter W Gray (PG 4001).*

Below: The hasty destruction also included the small engine shed, the rubble of which can be seen to the left of the now departing coal train. Although Brimscombe station is now long gone, the public foot crossing still exists to this day. *Peter W Gray (PG 4002).*

Above: The compact Brimscombe engine shed is seen in happier times in September 1964, with Collett 61xx 2-6-2T No. 6106 residing on the single track access road. Classified as a sub-shed of Gloucester (85B), the rudimentary stone structure was surmounted by a water tank which gravity fed the column, as well as an inspection pit and coaling platform. All to service the resident banking engine, which assisted trains up the grade to Sapperton. The May 1931 built locomotive was allocated to Swindon shed at this time and was to survive in traffic until December 1965, before passing into preservation in working order. Owned by the Great Western Society, it currently resides on static display at the Didcot Railway Centre. *David Christie (ANSTR.COM).*

Left: A bucolic vista is augmented by No. 1458 departing Brimscombe with a pair of autotrailers forming the 3.08pm train to Chalford. This particular autotank had a very short association with the line, arriving at Gloucester as late as July 1964 and being withdrawn only four months later.

The image foreground is dominated by the Thames and Severn Canal, which had been derelict since the 1920s, to become a haven for wildlife. As can be seen, fishing was a popular pastime, with eels to be found in abundance. The towpath also provided a wonderful walk along the stretch all the way from Sapperton Canal Tunnel to the Severn Vale.
Peter W Gray (PG 4000).

Left: St Mary's Crossing Halt was opened on 12 October 1903, with the two wooden platforms just long enough to handle the new railmotor service. The footbridge was added later, after a young woman was struck and killed by a train whilst using the road crossing. The halt was closed with the others on the line at the end of October 1964.

The chimney of St Mary's Mill can be seen to the left, sighted adjacent to the River Frome and Thames & Severn Canal. The gravelled access road to which was the sole reason for the gated crossing. This scene dates from the early 1960s, when all of the infrastructure was still in situ.

Lens of Sutton Association.

Above: By the time of this photo on 6 August 1965, the halt had been swept away leaving only the bridge and St Mary's Crossing Signal Box. Opened in 1875 to a GWR small square Type 2 design, it survives to this day and was Grade II listed in 1988. *Bernard Mills.*

Right: The cabin at St Mary's Crossing was really only a manned crossing ground frame, and not believed to have ever been a block post. It did however contain a small lever frame to work the pair of semaphore signals that protected the crossing, along with the associated block instrument repeaters.

This standard Reading Signal Works designed ground frame cabinet seen here, contained the crossing gate apparatus (signal repeaters, train approaching indicators and Annetts key release instruments) and is believed to have been mounted adjacent to the rear external wall of the cabin following introduction of colour light signalling on the line in October 1970.
Author.

Above: Looking in the direction of Brimscombe and Stroud, lined green liveried Collett 0-4-2T No.1444 approaches the crossing with a Gloucester to Chalford auto working in the Summer of 1964.
Bernard Mills collection.

Above: Taken from the elevated vantage point on the footbridge, the 'Large Prairie' Brimscombe banker assists a military train at St Mary's Crossing, circa 1956. The conflat wagons are loaded with Royal Ordnance 25-pounder field guns. These 3.45 inch howitzers were the British Army's main artillery field piece well into the 1960s. *George Heiron (1248).*

Below: The next main station on the line was at Chalford, which opened to passengers and goods on 2 August 1897. This panoramic view of the village and canal shows 70' autotrailer No. 6 and a train of 6-wheel coaches in the platform. *Lens of Sutton Association.*

Above: One of the original 1903 built pair of flat-fronted steam railmotors to be employed on the line, poses for the camera at Chalford prior to heading back down the valley. By the time of this circa 1905 view, it had received minor modifications to the front windows. Both vehicles had been withdrawn by the end of 1914. *Lens of Sutton Association.*

Below: A number of the new Collett designed 48xx Class 0-4-2Ts went to Gloucester shed in 1932, specifically for working the Chalford autotrains. An example is seen here with a brace of Churchward designed trailers shortly after introduction. *Lens of Sutton Association.*

Left: The well kept station buildings at Chalford were built in the familiar GWR style of red brick with blue engineering brick edging. Noted for its cleanliness, fresh paintwork, mats in every doorway and colourful flower borders; even in the latter years standards are seen to be maintained. In the distance can be glimpsed Chalford Signal Box, a 21-frame GWR type 7A design dating from 1897. The adjacent siding allowed the auto services to clear the main lines to lay-over and take water between turns. It also originally featured a Railmotor Shed, that was to be destroyed by fire on 8 January 1916, along with resident motor No. 48. This photograph is believed to date from the summer of 1963.
Bernard Mills collection.

Above: Prairie tank No. 6113 at Chalford on 24 October 1964. The engine out-stationed at Brimscombe shed was utilised on multiple roles from local passenger services, as the banker, and for shunting trip freight workings. The yard to the left had closed to goods traffic on 12 August 1963, and as can be seen the track had already been lifted, leaving an isolated loading gauge. *Peter W Gray (PG 3996).*

A pair of images captured in the first half of the 1950s, just to the east of Chalford station at the start of the 1 in 60 Sapperton Bank. Churchward 2-6-0 No. 6370 works a short unfitted Up freight which does not require the assistance of a banker, whilst an unidentified 'Castle' powers the nine-coach 11.45am Cheltenham-Paddington express up the grade. *George Heiron.*

Top: In the high Summer of 1956, Hawksworth 'Modified Hall' No. 7903 *Foremarke Hall* ascends Sapperton Bank and approaches the western portal of Sapperton Tunnel with the 3.55 pm 'Cheltenham Spa Express' to Paddington. In contrast to the previous view, this train comprises BR Mk.1 stock, except for the sixth coach which is a Diagram H33 Cross Country Composite Diner. *George Heiron (1280).*

Bottom: We conclude this pictorial journey up the Golden Valley with a scene that encapsulates the very essence of the railway in this beautiful part of Gloucestershire. On an August afternoon in 1953, three young ladies watch the progress of No. 1424 as it races its pair of pre-war autotrailers towards Chalford. The location is close to St Mary's Crossing Halt. *George Heiron.*

Model Railway News ran from 1925 to 1971 and was published by Percival Marshall. George Heiron's 'Sapperton Bank' photograph appeared on the November 1956 issue cover **(above)**. It was an inspiration for the author to take up railway photography. However, a design slip-up was made in making the far hillside horizontal, and as a result the train is leaning.

Reference: Gardiner SJ and Padin LC, *Stroud Road and Rail in Old Photographs*, Alan Sutton, 1987.

THE PANNIER STORY

PART 1 - BEGINNINGS

In this new benchmark *Western Times* serialisation, we intend to chronicle the history of that cornerstone of Great Western locomotive design, the ubiquitous Pannier Tank. The multi-part study will explore and question the complex history of the numerous classes.

The late 19th Century witnessed various experiments at Swindon in exploration of viable ways to produce more powerful locomotives. A small group of four-coupled tank locomotives formed part of this important programme:

Locomotive Number	Wheel Arrangement	Details
1	4-4-0T	Built 1880 with double frames, rebuilt as a double-framed 2-4-0T, withdrawn 1924.
13	2-4-2BWT	Built 1886, rebuilt as 4-4-0ST 1897, withdrawn 1926.
34 & 35	0-4-2ST	Built 1890, rebuilt 1895 as 0-4-4Ts, withdrawn 1908/ 6 respectively.
1490	4-4-0PT	Double-framed, built 1898, sold 1907.

The rationale for locomotives Nos. 1/ 13/ 34/ 35 is obscure, which is why they are earmarked for future consideration in a continuing programme to probe under researched aspects of the GWR's experimental locomotive endeavours. In the writer's opinion, William Dean was a courageous and innovative individual who deployed his creativity in search for ways in which future motive power demands, measured in speed capacity and power output, could be satisfied. So far as this quartet was concerned, it seems that they yielded no lasting impact upon GWR practice, but No. 1490 differed markedly by virtue of the 'P' in its designated wheel arrangement.

Above: Dean 4-4-0ST No. 13, as rebuilt from 2-4-2BWT in 1897, standing in the Brixham bay at Churston circa 1906. The first vehicle is a Diagram S7 Tadpole, a 4-wheel fish truck with planked ends and doors, and barred sides. The coach rake appears to comprise elliptical roofed 4-wheel Brake Composites at either end with a Clerestory All Third in between. The scene provides period interest in the baulk road, the home signal set between the running lines, the gas lamps and station fencing.

4-4-0PT No. 1490

Investigative work in the 1890s was directed along traditionalist lines in seeking ways to reconcile size increases with the four-coupled chassis and No. 1490 was at the heart of this exercise. The complex and hazard-fraught story of the forty locomotives generically termed 'Class 3521' (*Western Times*, Issue 8) highlighted the challenges and ultimately sealed the fate for larger tank engines with leading four-coupled driving wheels.

Pursuit of design alternatives was evident in No.1490 which ironically entered service during the month of the Penryn disaster that ended the career of Class 3521 as tank engines. The intention behind this locomotive was to find a more powerful replacement for the Metro 2-4-0T where acceptable stability might be possible through use of a leading bogie. The design was conventional for the time in having outside frames throughout and Mansell bogie wheels but the absence of a rear carrying axle limited room for cab and bunker, as with the preceding Metros. Leaf springs were used on bogie and leading driving axles but the space factor imposed helical springs for the rear axle.

Above: No. 1490 is pictured at Bath apparently working as station pilot or as a shunter, which was typical of this locomotive's duties later in its career on the GWR.

However, the main interest lay in the boiler as this was a period when the Dean/ Churchward partnership was exploring fresh ideas that would greatly benefit 20th Century locomotive practice. The first boilers with Belpaire fireboxes on the GWR comprised a batch of thirty of Type BR4 built between December 1897 and January 1900. Nineteen of the twenty-strong 4-4-0 'Badminton' Class carried this boiler from new. As a front-line express passenger type, this was a logical means of proving the relative merits of the new firebox form while the remaining eleven BR4 boilers were more widely dispersed. Five were quite logically fitted to smaller-wheeled 4-4-0s of the Duke Class when built in 1898/ 9 (original Nos. 3312/ 28-31) and five were fitted to six-wheeled tender locomotives:

Wheel Arrangement	No.	Class	Built	BR4 Fitted	BR4 Removed	Wdn.
2-2-2	160	157	Nov 1879	Jan 1900	Withdrawal	Sep 1905
2-2-2	1000	Queen	Mar 1875	Dec 1899	Withdrawal	Sep 1906
2-2-2	1125	Queen	Jun 1875	Oct 1899	Withdrawal	Oct 1906
2-4-0	3211	Barnum	May 1889	Sep 1899	Mar 1906	Nov 1934
2-4-0	3219	Barnum	Oct 1889	Nov 1899	Feb 1904	Jan 1935

Distribution of boilers around the turn of the century is a bewildering subject and the equipping these engines *might* have been rationalised through an investigation into how this type might fare on smaller machines. Nevertheless, it seems wasteful to have treated smaller single drivers with limited life expectancy in this fashion. The case for the Barnum Class might have been stronger but it appears that Swindon had acquired a batch of brand new boilers, some of which lacked an immediate purpose. However, the remaining BR4 vessel found a home with No. 1490 apparently through specific intent rather than convenience.

The BR4 type had a parallel domed barrel but was unusual in that the horizontal firebox 'roof' was 8.5" higher than the upper line of the boiler barrel i.e. significantly more than with later versions that were combined with either parallel or tapered barrels. However that carried by No. 1490 was an experimental version with the unusual feature that the firebox was wider than it was long. The purpose was to allow it to rest on the frames and to provide space for access to three 2" copper longitudinal circulating water tubes installed on either side; these were removed at an undisclosed date. The grate was steeply sloped with the brick arch supported by a pair of 2 7/8" arch tubes and the leading dimensions as built were:

Boiler Type	BR4
Barrel	
- length	10' 8"
- outside diameter	4' 7"
- pitch	7' 3.75"
Firebox Casing	
- length	4' 6"
- width	5' 10"
Firebox (inside)	
- length	3' 9.5"
- width [top]	4' 11.5"
- width [bottom]	3' 11.5"
Tubes	294 x 1 5/8"
Heating Surfaces	
- tubes	1392
- firebox	93
Grate (sq ft)	20.4
Boiler Pressure (lb/ sq in)	165
Cylinders	15.5" x 28"
Tractive Effort (lb - 85%)	15,785
Wheel Diameters	
- bogie	2' 8"
- driving	4' 7.5"
Wheelbase	5' 6" + 7' 0" + 9' 0"
Weight (tons)	
- total	50.8
- adhesive	34.8
- maximum axle loading	17.6
Water Capacity (gallons)	1075

These dimensions are believed to be correct as built but some time before December 1904, the number of tubes was reduced to 286, and thus the related heating surface to 1336 sq ft. There are records of later dimensional changes but these are believed to have been effected following its sale.

This engine failed to prove a viable prototype to replace the 2-4-0Ts (the largest version of which weighed 41.35 tons with a maximum axle loading of 15.25 tons) on grounds that the overall weight was excessive for use over the Metropolitan lines. It also proved too unstable for passenger duties which conclusion added to the empirical evidence already gathered through the adventures with Class 3521. Those tribulations had proven the wisdom of carrying wheels fore and aft with larger passenger tank locomotives working at higher speeds. Having been tried and found wanting in the Bristol and Gloucester districts, No. 1490 ended its career as a shunter at Swindon before sale in late 1907 into industrial service for which it seems to have been poorly suited. It was owned for about a year by Ebbw Vale Steel, Iron & Coal Co before sale to the Brecon & Merthyr Railway which was temporarily short of motive power. As that company's No. 35 it was largely used on branch passenger services before sale in March 1916 to Cramlington Colliery Co in Northumberland. There, several modifications were effected which apparently included frame strengthening. These changes of ownership suggest that there were other design problems, perhaps related to overall weight and axle loading. This unusual locomotive was broken up in 1929.

Concerning the boiler, early experiments with Belpaire fireboxes and with the parallel/ tapered barrels engendered complexity in different outer forms and in heating surfaces. Fitting the BR4 type boiler to No. 1490

Above: This side view of No. 1490 shows the limited space at the rear end. No record of the fuel capacity has been found but it looks to have been about two tons which would have inhibited use on longer distance work which might have been inferred from the locomotive's size. An effort appears to have been made to increase the capacity by fitting plating behind the bunker coal rails. ***GWR Official.***

was unique treatment for a tank engine suggesting an intended role as a test bed for other types. Beyond the minor changes to the standard tubes and removal of the water tubes, no further options were pursued under GWR ownership. Possibly, it was judged that the weight factor made further boiler modifications pointless.

To aid the originally intended test bed function, perhaps the novel means of water storage was devised to:

- facilitate complete boiler exchanges involving alternative barrel diameters and firebox profiles.
- assist inspection of the inside motion.
- ease servicing of the firebox water tubes.

No evidence has been traced to support this contention but it would have been logical to fabricate water tanks to be hung pannier fashion to help with these investigations. Conventional side tanks would have impeded maintenance while a saddle tank would have required careful fashioning of the inner profile to fit snugly over the firebox. Further experiments with alternative barrel/ firebox forms might have necessitated re-shaping of the saddle tank's inner profile with each change. Also, the water mass in a saddle tank raised the centre of gravity, increasing the potential for instability at speed.

Another argument in favour of the pannier arrangement which, despite extending only from the cab to the rear of the smokebox, was that the water capacity was commensurate with contemporary combined side/ back tank volumes. It seems possible that these factors converged to realise a viable compromise that would spawn one of the most enduring features of the Great Western's tank engine family.

0-6-4 (Crane) PTs Nos. 16 - 18

In 1876, three broad gauge 2-4-0Ts were under construction in the South Devon Railway works at Newton Abbot at time of absorption by the GWR. They were removed to Swindon for completion as standard gauge engines for branch line work and numbered 1298-1300 in the process. In 1881, No. 1299 was converted to a crane tank and initially used at Reading before transfer to Swindon at an unknown date. As evident in the illustration, the hand-operated crane could be rotated but there was no means of raising or lowering the jib which had a fixed radius of 16' 6" and the capability to lift 1½ tons maximum. This modest capacity restricted operational scope which was further limited by the jib exceeding the loading gauge. The ensemble was thus prohibited from entering confined spaces e.g. workshops. Also, movement over extended distances between work sites required disassembly/ re-assembly of the crane's key components which highlighted the design challenges generic to crane tank engines.

Above: Crane tank locomotives, which were few in number, presented particular design challenges as evident in this adaptation of 2-4-0T No. 1299. It was withdrawn in 1936 and broken up the following year, which shows that this must have been one of Dick Riley's earliest photographs of a GWR subject. ***R C Riley.***

Around 1900, the need to improve materials handling in engineering workshops necessitated mobile cranes of greater capacity and broader availability. No. 1299's limitations had shown the need for an alternative approach and in April 1901, two crane tanks were introduced that provided a more expedient solution. They were based on the standard 0-6-0ST Class 850 but with the frames extended at the rear to carry a platform on which the crane was mounted. The platform and frame extensions were supported by a bogie resulting in an 0-6-4 wheel arrangement.

Class 850 was a good starting point for this exercise as the smallest design of 0-6-0ST, with lively performance and modest axle loading. The crane was operated by steam drawn from the locomotive's boiler. One vertical cylinder drove the mechanism that rotated the crane through 360 degrees while another operated the jib. Provision of a large counterweight allowed a significant increase in load capacity compared with that of No. 1299:- maximum six tons on an 18' radius and nine tons on a 12' radius. When not in use, the jib lay longitudinally in the forward position across the roof of the open-backed cab and rested on a transverse support bracket mounted at the leading end of the firebox. In this position the jib and counterweight resided in line with the top of the chimney and so within the loading gauge.

Stowing the jib required special treatment for the superstructure. Two standard Group 5 boilers (then typically fitted to 0-6-0ST Class 850) were built with customary raised fireboxes but domeless. The result was Type R0, an old design last used in 1864/ 5 for Class 302, the first 0-6-0STs built at Wolverhampton. However, this alone was insufficient as a saddle tank would still encroach upon the space needed for the jib while side tanks would have restricted access to the motion. The solution was installation of pannier tanks as used three years earlier with 4-4-0PT No 1490.

The same principles for mounting and support of the tanks were adopted, the only difference being that these stretched from the smokebox front to the forward face of cab. By this route, the GWR acquired its next two pannier locomotives, albeit in specialised form and retaining the round-topped boiler. Numbered 17 and 18, the pair were named *Cyclops* and *Steropes* respectively. (In classical mythology, the Cyclopes were a race of giant one-eyed creatures renowned as craftsman and for their building skills). The leading dimensions were:

Boiler Type	R0
Barrel	
- length	10' 0"
- outside diameter	3' 10"
- pitch	6' 1"
Firebox Casing	
- length	4' 0"
Tubes	214 x 1 3/4"
Heating Surfaces	
- tubes	1007
- firebox	80
Grate (sq ft)	12.6
Boiler Pressure (lb/ sq in)	150
Cylinders	16" x 24"
Tractive Effort (lb - 85%)	15,820
Wheel Diameters	
- driving	4' 1.5"
- bogie	2' 8"
Wheelbase	7' 4" + 6' 4" + 6' 6" + 6' 0"
Weight (tons)	
- total	63.6
- adhesive	39.6
- maximum axle loading	14
Water Capacity (gallons)	1200

Twenty years later, another crane tank was required. The urgency of the need for this machine is unclear as it was completed in April 1921 but entered service 12 months later as No. 16 *Hercules* (Roman equivalent of the Greek divine hero Heracles, son of Jupiter and famous for his strength). It was formally entered into running stock on 29 December 1923. By this time the type B4 (Belpaire) boiler was in widespread use with Class 850 so a unique domeless version (type BR0) was installed to accommodate the jib. Differing dimensions with this boiler:

Boiler Type	BR0
Firebox Casing	
- length	3' 3"
- width	3' 5"
- height	5' 5.5"
Tubes	207 x 1 5/8"
Heating Surfaces	
- tubes	904
- firebox	76
Grate (sq ft)	11.2
Boiler Pressure (lb/ sq in)	165
Tractive Effort (lb - 85%)	17,410
Water Capacity (gallons)	1200

The trio was mainly associated with Swindon although No. 17 was at Stafford Road works for around 15 years. They were also used to help with civil engineering work on lineside structures. By the late-1920s, more modern forms of materials handling had been implemented and their usefulness started to wane. All three were at Swindon in 1933, the year in which they ceased work, remaining there until withdrawal in September 1936, and broken up around a year later.

Above: Based on a Class 850 0-6-0ST, No. 17 *Cyclops* was a more practicable form of crane tank where pannier tanks for water storage enabled the jib to be stowed within the loading gauge. The location is Wolverhampton, Stafford Road circa 1910.

Below: The flexibility of the 0-6-4PT crane engine is demonstrated here with No. 18 *Steropes* (in mythology one of the Cyclopes) engaged in civil engineering work on the platform faces at Paddington in 1933.

The Belpaire Firebox

Integral to the requirement for tanks in pannier form was the Great Western's enthusiastic adoption of the firebox invented by Alfred Belpaire of Belgium in 1864, although it was about 1882 before the principle started to gather broader acceptance. RPC Sanderson, an Englishman who had qualified as an engineer in Germany, was working for the Shenandoah Valley Railroad, USA in 1875 when he heard about this type of firebox. Through a colleague from his college days who by then was working for Henschel, he obtained a sketch drawing of the new style. Sanderson showed this to C Blackwell, Motive Power Superintendent of the Shenandoah who recognised the potential. Orders were placed for passenger and freight locomotives with manufacturers Baldwin of Philadelphia and Grant Locomotive Works of New Jersey. The Belpaire firebox then started to gain partial recognition in the United States.

Around the turn of the century, the Pennsylvania Railroad (one of the comparatively few major U.S. operators to build its own motive power) took up the principle. 'Pennsy' was well regarded for its scientific approach to motive power construction which included its pioneering stationary testing plant at its Altoona workshops which was the inspiration for Churchward's facility at Swindon.

Pennsy wholeheartedly embraced Alfred Belpaire's invention as prominently demonstrated in its express passenger pacifics. Introduced in 1914, Class K4s was in production until 1928 by which time there were 425 in service. These remarkable machines handled all the railroad's premier passenger duties until arrival of 4-4-4-4 Duplex Class T1 in the 1940s, and continued in front line work until the end of PRR steam.

Despite this example, acceptance in the USA was far from universal. The round-topped boiler remained widely popular, partly attributed to US railroads' reliance on commercial manufacturers and the resultant lower manufacturing cost which would have been attractive in refining sales prices. The round-topped style assumed its most pronounced form through the vast Wootten firebox whose width at the base stretched to the limit of the loading gauge. These vessels burned low grade coal, even slag, and operated with great economy. Produced before mechanical stokers were invented, the Wootten firebox demanded special firing methods as with the Erie Railroad's 0-8-8-0 Camelback Compound Mallet which was equipped with two fire doors, served by a pair of stokers. In fuel quality, form and proportion this was the antithesis of Swindon's Belpaire philosophy.

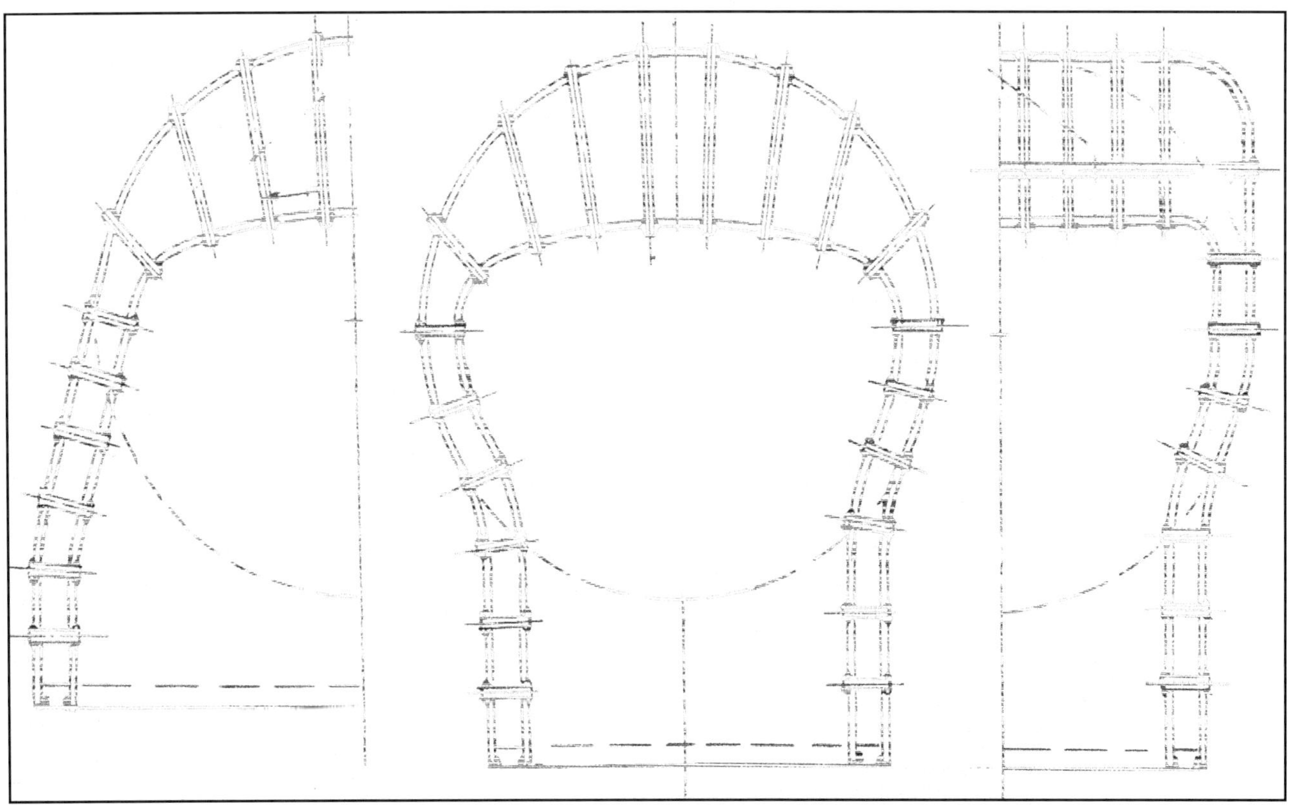

Above: Cross-sectional drawings of (left to right) Wootten, Round-Topped and Belpaire fireboxes. The latter also depicts transverse crown stays. ***Alan J Haigh.***

WESTERN TIMES

In these islands, rather curiously the first locomotive to carry a Belpaire boiler was 3' 0" gauge Schull & Skibbereen Light Railway 4-4-0T No. 4 *Erin* supplied new in 1888 by Nasmyth, Wilson & Co. Most of this 15-mile system was a twisting, hilly roadside tramway in Co Cork, Ireland and this locomotive was a great improvement over the company's three puny 0-4-0T engines in use from the line's opening two years earlier. However, in this case it can be assumed that use of a Belpaire boiler was accidental rather than to optimise performance!

Among Churchward's many strengths was his willingness to look outside Britain for advances in locomotive design and he recognised the advantages in the Belpaire firebox as a means of maximising the benefits offered by the higher calorific value of Welsh coal. UK-based opponents of the Belpaire style, most commonly associated with the Doncaster school, held that advantages were off-set by the greater cost in manufacturing the upper part of the vessel's distinctive angular shape. However, the Belpaire's inner firebox broadly followed the contours of the outer shell which enabled the stays to be of regular length and set at right angles to the inner firebox walls. The stays at the crown were typically longer than those at the sides thereby allowing increase in the volume of water above the firebox and at the throatplate, which are the hottest parts of a boiler.

Round-topped and Wootten fireboxes require stays of differing lengths which must be set at varying angles, thereby making the process of stay production and installation more complex. Citing the US example, capital cost advantage was probably off-set by subsequent greater maintenance expenses.

The GWR was notable for adoption of the Belpaire firebox across a range of types from express tender to small tank locomotives. It was almost universally applied by Swindon in new construction from the early 1900s and in replacement of life-expired round-topped fireboxes on older engines.

Above: A G Watson, CME of South African Railways [1929-36] in reaction to a confused investment programme in the 1920s disliked articulated locomotives and developed a rigid-framed fleet that used the Wootten boiler. South African coal was below the quality of Welsh but better than much of the fuel used by the Wootten type in the USA. SAR 4-6-2 Class 16DA was the first to use this type of firebox on SAR and its steaming capacity was regarded as phenomenal with a slow rate of combustion and minimal emission of unburned material. This view of No. 879 shows how the firebox width made full use of the loading gauge.

The Six-Wheeled, Six-Coupled Tank Engine Fleet

The importance of this group to the GWR is best measured from 31 December 1892, i.e. the first year end for the company following consolidation of a fully integrated narrow (standard) gauge system. The composition of the motive power fleet at that date:

Wheel Arrangement	Number of Locomotives	% of Fleet Total
2-4-0	182	
2-2-2	101	
0-6-0	565	
Sub total	**848**	**41.4**
4-4-0T	8	
2-4-2T	1	
2-4-0T	126	
0-6-0T	**842**	**41.1**
0-4-4T	51	
0-4-2T	159	
0-4-0T	14	
Sub total	**1201**	
Fleet total	**2049**	

The continuing dominance of the 0-6-0T wheel arrangement was apparent in the following years:

Year (31 December)	Fleet Total	0-6-0T Total	0-6-0T % of Total
1897	2393	943	39.4
1907	2992	1119	37.4
1917	3127	1105	35.3
1927	4007	1195	29.8
1937	3632	1126	31
1947	3856	1251	32.4

(Even in the nationalised era this pattern was sustained. It was 1960 before the 0-6-0T population dropped to triple figures when 972 locomotives comprised 37.2% of BR(W)'s steam fleet).

These figures show the importance of the 0-6-0 tank locomotive around the turn of the century. Thus the effectiveness of the Belpaire firebox combined with plentiful supplies of Welsh coal had a major impact upon operating economics. Acceptance of the new firebox would have been straightforward except that most of the 0-6-0Ts then in service were fitted with saddle tanks.

Wolverhampton-built 0-6-0ST Class 1501 No. 1532 as built in September 1879. Typical of saddle tanks in early condition, the bunker had modest coal capacity but over the years this feature would assume different forms of enlargement both before and after acquisition of panniers. The northern works applied detail differences from Swindon as apparent in the roll-top style chimney and the more arched profile of the half-cab roof. This locomotive received a Type B4 Belpaire boiler plus pannier tanks in September 1921 and remained in service until July 1948. ***Stephenson Locomotive Society Collection.***

A design challenge with the tank locomotive type is that everything must be crammed onto a single frame structure. This is desirable with a smaller engine in enhancing adhesive qualities but less beneficial in restricting access to internal mechanical parts. The saddle tank provided a neat compromise, so long as the firebox was flush with the boiler barrel.

This article commenced with reference to William Dean's five tank engine oddities and his penchant for experimentation which yielded a valuable store of empirical information on good ideas and equally on those better avoided; knowledge which Churchward would exploit with profound effect. Of all those essays, Dean's solution to the Belpaire/ water storage conundrum as devised with 4-4-0PT No. 1490 probably bestowed greater value upon the company's motive power saga than any other of his individual initiatives. Realisation of that advantage would become a cumbersome process deriving from the company's uniquely complex motive power situation in the final quarter of the 19th Century.

The Gauge Legacy

Manufacture of 0-6-0 standard gauge saddle tanks started at Wolverhampton in 1864 with a remarkable number produced up to the early years of the 20th Century. Parallel construction started six years later at Swindon with chronological order of introduction for the classes at both works as follows:

WOLVERHAMPTON					SWINDON				
Class	**Built**	**Frames**	**Driving Wheel**	**Number**	**Class**	**Built**	**Frames**	**Driving Wheel**	**Number**
302	1864/5	D	4' 6"	8	**1076**	1870-81	D	4' 6"	266
1016	1867-71	D	4' 6"	60	**1813**	1882-84	IF	4' 6"	40
645	1872/3	IF	4' 6"	30	**1661**	1886/7	D	5' 0"	40
850	1874-95	IF	4' 0"	170	**1854**	1890-95	IF	4' 6"	120
119	1878-83	SP	4' 7.5"	11	**2721**	1897-1901	IF	4' 7.5"	80
322	1878-85	D	5' 0"	6					
1501	1878	IF	4' 6"	72					
655	1892-97	IF	4' 6"	52					
2021	1897-1905	IF	4' 1.5"	140					
Total Builds				**549**	**Total Builds**				**546**
Number of Years of Building				41	**Number of Years of Building**				31
Annual Average Number of Builds				13.4	**Annual Average Number of Builds**				17.6

Key to Frame Type: D = double; SP = sandwich outside & plate inside; IF = inside only.

Right: Small 0-6-0STs of classes 850 and 2021 were a Wolverhampton speciality. Their diminutive size, low axle loading and lively performance made them handy for deployment throughout the more obscure corners of the GWR network. Class 850 No. 1925, seen here at Didcot, was the first member of Lot O2 introduced in 1883. Together with sister engine No. 2007 (Lot Y2 of 1891/ 2), they achieved celebrity status as the last pre-Grouping vintage saddle tanks (excluding the much younger Class 1361) to remain in service. No. 1925 was withdrawn in April 1951 and No. 2005 in December 1949.

Stephenson Locomotive Society Collection.

Above: At one time, the Buffalos together with the Armstrong 0-6-0s worked as the company's prime haulers of long distance goods traffic before their deployment on a diverse range of other duties. No. 1601 [Lot 50 of 1879] was in near original condition although it is thought that this batch entered service without the bunker coal rails. Arguably the most popular of the pre-Grouping 0-6-0Ts, this photograph displays much contemporary detail that remained following conversion to pannier tank status. It also reveals a drawback with the saddle tank configuration as access to the filler required some agility on the part of the fireman.
Stephenson Locomotive Society Collection.

The differing strands of design and construction stemmed in large part from the origins of the two facilities. Swindon works had formed an integral part of the GWR's organic development from early days. On the other hand, the works at Wolverhampton Stafford Road were rudimentary when they came into the fold through acquisition of the Shrewsbury & Birmingham Railway on 1 September 1854. Joseph Armstrong had been in charge of the S&BR's motive power from 1853 and on becoming the GWR's Northern Division Locomotive Superintendent, he developed Stafford Road from a service facility for contractor-sourced locomotives into a fully-fledged works that built narrow gauge locomotives. Narrow Gauge construction started at Swindon in 1855 but that works' main focus remained with maintenance of broad gauge services, and in development of the convertible fleet needed for smooth gauge transition in the southern segment of the network.

Differences between the two centres were accentuated by complications in management succession for the entire system. Joseph Armstrong moved to Swindon in 1864 as Locomotive Superintendent to succeed Daniel Gooch who had resigned to work on the Transatlantic Cable project. He was followed at Wolverhampton by George, his younger brother by six years. Joseph proceeded to modernise Swindon works and with a heavy workload, he brought in William Dean from Wolverhampton. The company was evidently short of competent managers as Dean had proved his mettle at Wolverhampton by concurrently occupying three positions (Chief Assistant of the Locomotive Superintendent, Works Manager and Chief Clerk). In 1868 at Swindon, he became Chief Assistant to Joseph Armstrong, an appointment which George took as an affront.

Joseph in his Victorian moral rectitude was determined to avoid suspicions of nepotism that could have attended George's advancement as his assistant. Further, Dean's promotion at Swindon was his de facto anointing as Joseph's successor which in the normal course would have been occurred in say 10 years' time. However, Joseph passed away in 1877 at the age of 60 years, apparently worn out by overwork, and Dean became the company's Locomotive Superintendent, initially on a trial basis. George, then aged 55 years and already disgruntled at having been passed over the first time objected to the expectation that he must now report to Dean (then aged 37 years), his erstwhile subordinate at Wolverhampton. He is quoted as declaring "...he didn't give a damn for any man and was taking orders from none. He only gave orders." (Harry Holcroft, *The Armstrongs of the Great Western*, 1953).

Although an equally strong-minded individual, Dean handled this delicate situation by leaving George Armstrong largely to his own devices and he remained in charge at Stafford Road until retirement at age 75 years in 1897 (ten years after his scheduled retirement age). New construction at Wolverhampton was predominantly focussed on tank locomotives and the works continued to express its individuality in key features like frame structure and in detail matters e.g. livery and chimney styles. Policy differences were particularly evident in the wheel arrangements of two numerically large families of four-coupled tank engines: 0-4-2T Class 517 from Wolverhampton and 2-4-0T Metro Class from Swindon.

There was notable merit in some of the Wolverhampton initiatives. Single inside frames in new construction were adopted earlier than at Swindon. Exceptions were double-framed classes 119 (conversions of Class 79 0-6-0s built at Swindon in 1861/ 2) and 322 (conversions of 0-6-0s supplied by Beyer Peacock in 1864). The small 0-6-0STs of classes 850 and 2021 with their lightweight go-anywhere flexibility were particularly useful additions to the fleet.

In contrast, Swindon adhered for longer and in greater measure to double frames. Class 1076 (nicknamed 'Buffalo' for reasons that seem to have been lost in time) were a particularly successful and flexible class. However, Class 1661 was the product of a standardisation scheme inaugurated by Dean between 1884-7 that embraced four classes. This was probably the least successful design in the 19th Century 0-6-0ST construction programme.

The Broad Gauge imposed a financial and organisational burden upon the GWR that stretched well beyond operational difficulties such as the transhipment mayhem at Gloucester, or the heroic efforts during that frantic weekend in May 1892, or the loaded expense of creating a locomotive and rolling stock fleet with limited life expectancy. An additional burden was imposed through Wolverhampton's separatist tendencies which detracted from cohesive fleet management in motive power.

Acceptance

Although the inherent advantages of the pannier tank concept were proven with No. 1490 and the crane engines, introduction as a feature with general service locomotives was an extended affair and never completed. The early pattern of pannier tank and Belpaire boiler installation hardly reflected a focussed strategic approach and it was really only after about 1910 that the programme started to gain momentum.

Class 1813

This forty-strong 0-6-0T entered service between September 1882 and January 1884 and all but one (No. 1817) received saddle tanks, commencing in 1894. Built with the S0 two-ring domeless type boiler, 39 in due course graduated to the S2/ S4 versions. In July 1903 No. 1813 acquired short pannier tanks (in the style fitted to No. 1490) which were replaced by the full length version in November 1911, yet it was December 1918 before a Belpaire boiler was fitted. In November 1905, No. 1814 was fitted with short tanks (carried possibly until January 1919) and acquired a B4 boiler in September 1905 – the two fitments were probably carried out during the same workshops visit. Odd man out No. 1817 had a B4 boiler fitted in August 1904 and acquired pannier tanks the following December, never having carried either an S2/ S4 boiler or a saddle tank.

Above: Dean '1813' Class No. 1824 following its conversion to pannier tank form in May 1912. Built at Swindon as a side tank in December 1882, it was originally rebuilt with a S4 boiler and saddle tanks in April 1901. Withdrawal came in June 1938.

Class 2721

The final batch [built 1901] of five members (Nos. 2796-2800) of this 80-strong class were the first 0-6-0Ts built new with Belpaire Type B4 boilers. No. 2796 was unique among the pre-Grouping engines in always carrying this boiler type but it only received pannier tanks after three years' service. This seems to have been the point at which the awkward nature of Belpaire firebox/ saddle tank combination first emerged but full conversion of the remaining four was prolonged:

No.	Built	Later Boiler Types	Pannier
2796	Feb-01	-	Mar-04
2797	Mar-01	S4: Mar-07/ S2: Apr-09/ B4: Nov-11	Oct-11
2798	Mar-01	S4: Aug-09/ B4: Jan-28	Jan-28
2799	Mar-01	S2: Jul-05/ S4: Mar-10/ B4: Jun-16	Jan-25
2800	Mar-01	S4: Mar-09/ B4: May-18	Jul-18

Standard gauge design policy up to 1892 relied almost exclusively upon six-wheeled locomotives and in an era of comparatively light passenger trains and leisurely schedules, modestly-sized motive power had filled the need adequately. Two incidents alluded to in earlier issues of *Western Times* (Nos. 3 & 8) served notice that traffic growth would stretch existing boundaries for speed, size and power. The derailment of No. 3021 *Wigmore Castle* in September 1893 led to a high speed 4-2-2 with superb riding characteristics but restricted adhesive capacity. The fate of 0-4-4BT Class 3521 as a general passenger type was sealed with the Penryn accident in April 1895.

These experiences were extreme examples but nonetheless reflected the mixed fortunes in the latter part of the 19th Century that attended development exercises necessary to produce more advanced machines. The company's large fleet of sturdy, reliable tender and tank 0-6-0s would continue to bear the brunt of traffic demands in the interim. The tank engine family's work generated a large portion of operating revenue, and it was notable that on taking over locomotive affairs, Churchward only needed to pay minimal attention to this major component of the Company's fleet.

The Wolverhampton-built contingent will be discussed in **The Pannier Story: Part 2**.

Above: The photograph serves as a 'health warning' against drawing too many definite or sweeping conclusions about the diverse forms that the 0-6-0ST/ PT tribe assumed. As a very general rule, changes (most apparent in tanks, cabs and bunkers) followed a chronological sequence which was honoured in the breach by Class 2721 No. 2743. This was the last class of large saddle tanks to be introduced and this example built at Swindon in 1899 as part of Lot No. 115 has undergone 'modernisation'. The cab retains the original side sheets but has been enclosed with an early Collett-style roof extension and back sheet, and has been fitted with the last form of bunker. The photograph is undated but was probably taken around 1930. Pannier tanks were fitted as late as June 1933 yet a B4 type Belpaire boiler had been installed in February 1917. Despite these changes, unusual in their application counter to the normal pattern, this locomotive had a long and useful life being withdrawn in October 1950. ***Stephenson Locomotive Society Collection.***

SUNDAY 26 FEBRUARY 1939

FROM THE ARCHIVES OF R C RILEY

As an antidote to the diesel and colour debates of recent issues, a random date has been chosen from Dick Riley's archives to provide a flavour of the GWR scene before the War. It was only some years after the conflict, under the influence of the late great M W Earley, that Dick believed his skills to have fully matured but his innate craftsmanship nonetheless shows through in these examples. Many early photographs were taken in his native south London and they have their own particular fascination. A slightly fuzzy view of LBSCR Class A1X No. 2644 in front of a 20-ton Pole wagon in Stephenson & Clarke livery certainly struck a chord.

Fortunately, Dick soon turned his attention to the GWR and on that Sunday, he visited Didcot and then Swindon. A feature of his pre-war material was emphasis on static subjects, imposed by his camera's limitations. This meant that shed scenes are common and in some cases images were confined by structures and adjacent locomotives that limited possible angles, and imposed lighting difficulties in the darker corners. Nonetheless these views provide an attractive memento of a railway day out in the 1930s.

The R C Riley Collection forms an integral element of the *Western Times* endeavour for reasons beyond the inherent quality of the photographic work. Of the inventory of 18,000 images, about 40-50% relate to GWR and BR(W) subjects with inevitable emphasis on the latter. As with any database, its value is greatly advanced by the ease with which search and recovery can be conducted. Recorded information in all but a minority of cases covers date, location, and locomotive number which greatly helps the processes of selection and validation, the convenience of which cannot be underestimated by the editorial team.

Dick's first locomotive photograph was of LNER No. 4473 *Solario* which quickly proved the inadequacies of his Box Brownie. A Kodak folding camera acquired in 1937 was an improvement, but the 1/100 shutter speed precluded moving trains. After the war, a Zeiss Ikonta with a 1/250 speed allowed a wider range of subjects and improved results. In the 1950s he acquired an Agfa Isolette and then an Agfa Record. He asserted that he never paid more than £20 for a camera for black-and-white photography.

Above: Few of the subjects photographed were in steam or on the move but Class 56xx No. 5689 of Swindon depot was captured while shunting in Didcot shed yard. This engine was an immigrant from Wales between December 1930 and February 1962 (Westbury and Swindon) but was withdrawn from Radyr in June 1965. ***R C Riley (RCR 115).***

Above: Dean Goods No. 2393 was also in the yard, attached to a 3500-gallon tender which was younger than its companion, and bearing the art deco emblem that seemed out of place alongside an engine then over 49 years old. This locomotive was built as part of Lot No. 104 in 1890 and entered traffic in September of that year. About seven months after this photograph, it was the second oldest class member to be requisitioned, and was re-numbered 113 in the War Department series. By November it was deployed in France in support of the British Expeditionary Force. Taken prisoner in 1940, it became in due course Société Nationale des Chemins de fer Français No. 030W036. After the war the surviving French Dean Goods were in generally poor condition but this engine, after some attention in SNCF workshops, was sufficiently sound to form part of a group despatched to China under the auspices of the United Nations Relief and Rehabilitation Administration. The Chinese Civil War was then raging and it went 'missing in action'. *(RCR 117).*

Left: Even after World War Two, Didcot was a favoured home shed for veterans, most prominently 4-4-0 No. 3440. Another four-coupled resident that Sunday was Duke Class No. 3256 *Guinevere*. This locomotive was withdrawn in August 1939 and its boiler mounted on the chassis of unnamed Bulldog No. 3429 to form Earl or 32xx Class No. 3228, later No. 9028 and withdrawn September 1957. This was the last Duke/ Bulldog hybrid to be completed before the War terminated this exercise. The shadows cast by locomotives on the adjacent road to the right suggest that this was the only angle from which the Duke could be captured and a 'creative' view of a locomotive in tidy condition resulted. *(RCR 118).*

Above: A younger resident that day was 0-6-0PT Class 8750 No. 3608. Lot No. 314 comprised Nos. 3785-99/ 3600-34 built at Swindon between September 1938 and July 1939; this particular locomotive entered traffic earlier that month and was withdrawn from Southall in June 1965. *(RCR 120).*

Below: No. 6866 *Morfa Grange* was approaching completion in the erecting shop at Swindon. This locomotive entered traffic the following month and spent almost its entire career allocated to Tyseley from where it was withdrawn in May 1965. *(RCR 112).*

Right: Dick Riley was widely respected for his superb camera work in the 1950s and 60s, and the quality of some pre-war photographs was sub-standard in comparison. This view of Class 4575 No. 5539 awaiting repair has been included for its technical interest. It has suffered front-end damage, probably from a collision rather than a heavy shunt, considering the degree to which the footplate has been buckled.

Churchward's two-cylinder locomotives used conventional plate main frames that terminated just ahead of the leading coupled axles. The American style of cylinder construction required bar frame extensions which were bolted to the inside faces of the main frames. There was concern over resultant structural weakness which was alleviated from 1908 by the fitting of sturdy struts from the front of the bar frames to the smokebox, abutting the flanges of the saddle. This feature was considered only necessary with locomotives with a pony truck (plus one with a bogie) and smaller driving wheels where the lower cylinder height necessitated a downward set of the bar frames. With No. 5539, the struts have fulfilled their intended role by containing the extent of the damage. *(RCR 134).*

Above: Another workshop scene featured 2-8-0 Class 30xx (ROD) No. 3020, temporarily masquerading as an 0-4-0. Without delving too deeply into the complex relationship between the GWR and this type, the engine portrayed was the third example to carry this GWR running number. It had formed part of Railway Operating Division Order No. L707 placed with North British Locomotive Co, Atlas works, Glasgow [makers No. 22088] in August 1918 and was built in 1919 as ROD No. 1794. Completed too late for service in France, it went straight into short-term store before becoming one of 93 locomotives on loan by the Government to the Great Central Railway between 1919 and 1921. On termination of that arrangement it returned to store, this time at Royds Green near Leeds and became one of eighty purchased for £1,500 apiece in May 1925. It was temporarily numbered 3050 but following assessment as suitable for long term service, it was re-numbered again as 3020 and extensively 'Westernised'. This process encompassed fitting of MB type boiler, top feed, Swindon style superheater, copper firebox and GWR safety valve bonnet. No. 3020 returned to service in re-conditioned form in May 1926, as one of the first two so treated. Many received GWR-style smokebox doors but not this example which has retained its classic Robinson smokebox door but has lost the stylish original chimney in favour of what looks like a 47xx type. This engine survived on the books of Banbury until June 1954. *(RCR 137).*

Above: Class 8750 No. 9757 to the left was built at Swindon in June 1935 (withdrawn from Swansea East Dock in August 1962) whereas 2-8-2T No. 7250 was fresh out of the box having been completed in February 1939, as a rebuild of 2-8-0T Class 42xx No. 4219. The programme to broaden the operational range of the 2-8-0Ts by increasing the fuel capacity had been in hand since August 1934 and No. 7250 was the 51st so treated (withdrawn from Radyr in September 1964). Only three more were converted, the last being No. 7253 in December of that year. *(RCR 131).*

Below: The Collett 48xx Class attracted derision outside GW circles due its old-fashioned appearance and outmoded wheel arrangement, features natural enough in a type that clearly traced its ancestry back to Wolverhampton-built Class 517 of 1868. Setting aside the whims of those who valued styling over performance, these sprightly and economical locomotives filled an important niche in working branch and suburban auto trains. No. 4857 had been on the books of Stourbridge shed prior to entering the works on 29 December 1938. It returned to Stourbridge two days after Dick Riley's visit. This engine's closing years were filled with uncertainty as its last active allocation was Birkenhead in 1957, followed by periods in store or awaiting possible repair at Kidderminster and Oswestry until sale for scrap from Swindon on 21 September 1959. *(RCR 141).*

BOOK REVIEW

CORNERS OF THE CAMBRIAN

Compiled by
JEFFERY GRAYER

The Transport Treasury (ISBN 978 191325 1598)
Paperback, 112 pages, 273mm x 215mm
£14.95

It is not our usual practice at *Western Times* to review books that emanate from our own publishing stable, but on this occasion we have made an exception. This is partly due to an apparent shortage of new GWR themed titles released at the tail end of 2023/ early 2024, but is also justified by a desire to shed more light on the less popular corners of the empire. Dare I offer that the Cambrian falls into this category.

The Cambrian Railways, formed by the amalgamation of several different companies, became part of the GWR at the Grouping and after nationalisation part of the Western Region then, in 1963, the Midland Region of British Railways. It retained much of its rural charm by the very nature of the countryside through which it passed and by its use of often vintage motive power which attracted both tourists and enthusiasts alike. By the mid 1960s however, many of its routes and stations had been closed and rationalisation and standardisation had destroyed much of its attraction.

This latest album follows the now popular house-style, drawing upon photographs from within The Transport Treasury archive, selected and captioned by regular author Jeffery Grayer. It contains some 160 images captured between the 1930s and 1970s, two supporting maps and a timetable. Reproduction quality is overall very good, and it is nice to see that a UK-based print house has been used.

We open with homage paid to the principal train associated with the region, the 'Cambrian Coast Express', and are then taken on a tour of this once great 300-plus mile network. The important hubs of Oswestry, Welshpool, Machynlleth and Aberystwyth feature, but the small branches to Kerry, Dinas Mawddy, Llanfyllin and Llangynog also gain attention. Civil engineering masterpieces such as Barmouth Bridge and the Friog avalanche shelter are included, as is the junction with the Shropshire and Mongomeryshire Railway at Llanymynech. For those of a more mechanical persuasion, many locomotive classes associated with the lines appear, of course including the venerable Dukedogs and Manors.

The image selection includes some real gems, with the panoramic portrait by Roy Vincent of a Dukedog, Small Prairie double-header exiting Barmouth Tunnel on page 61, particularly catching the eye of the reviewer. For devotees of the brace of narrow gauge lines inherited by the Cambrian, namely the Welshpool & Llanfair and Vale of Rheidol, only three pages are featured. They do appear to be new images however, which tempers the initial disappointment.

Overall, this book provides a comprehensive and good value resume of the latter years of steam operations on the Cambrian, stretching from Whitchurch and Wrexham in the nort-east, to the shores of Cardigan Bay in the west, and south via the Mid Wales Line to Brecon. It is recommended as a nostalgic reminder of a once great web of routes, of which only 118 miles of the original main line remain today, between Buttington Junction to the north of Welshpool and the resorts of Aberystwyth and Pwllheli.

ASM.

FORTY PAIRS OF NUMBERPLATES

PART 2 - TENDER ENGINES

The various modifications to Nos. 3521 to 3560 in tank locomotive form effected up to November 1892 had achieved only marginal improvement in stability. Originally, standard gauge Nos 3521-40 were used on local and branch work in the London and Bristol divisions while the broad gauge engines were deployed beyond Newton Abbot on similar duties plus goods trains. While it is uncertain exactly when as 0-4-4BTs they first worked mainline passenger services, it would seem to have been soon after gauge conversion in 1892. The accident between Doublebois and Bodmin Road on 13 April 1895 involving Nos. 3521 and 3548 marked a fundamental change in the pattern of these locomotives' duties.

Their instability as tank locomotives required a radical solution and how Dean overcame the problem was probably helped by his diverse experience. As mentioned earlier, he had for a period held three senior executive positions concurrently at Stafford Road. That smaller and more intimate working environment would have allowed him a better appreciation of other departments' responsibilities. There is no evidence of Dean having been tutored in financial management but his role as Chief Clerk would have provided a useful insight into the importance of costing disciplines in the engineering function.

This background must have proved invaluable on moving to Swindon. The company was still recovering from near bankruptcy of a few years earlier while having to cope with the dual gauge incubus and its ultimate elimination. The late and much respected Joseph Armstrong had produced the interim solution through creation of the first Convertibles (0-6-0STs Nos. 1228-1237 as part of Lot No. 43). However, expansion of that specialised fleet with inherent short term life expectancy would have called for superior evaluation techniques and recognition of the need to conserve slim monetary resources.

Above: No. 3534 re-entered service as a 4-4-0 in December 1899 and retained its S2 boiler until replacement in July 1907 by a Type S4 version which was dimensionally similar but with the dome centrally placed. In April 1918, this in turn gave way to a Type B4 Belpaire boiler that was carried until withdrawal in October 1927. In this view, the locomotive's S2 boiler condition, including the glistening brass dome, is indicative of the amount of cleaning applied even to secondary motive power in those days. ***Real Photographs.***

The successful solution for the 0-4-4BTs' problems appeared radical as it involved conversion to the 4-4-0 wheel arrangement whereby the chassis was partially reversed relative to the boiler. A length was cut off the inside frames at the bogie end and a shorter length was welded to the inside frame at the cylinder end. The inside frames could not be reversed as they were joggled inwards behind the firebox. This process reduced the overall locomotive length by a net 2' 5" and the inside frames were cut away below the cylinders to accommodate the bogie in its new position. The outside sandwich frames were shortened to match the length of their inside companions, and were reversed. The distinctive pattern of the outside frames around the driving axles was thus retained from their days as tank locomotives. What at first sight seemed a major challenge actually proved elegantly simple and cost effective. The frame layout of all forty locomotives was modified in this fashion and as 4-4-0s, they proved capable with working careers notably less eventful than in their earlier years

These 'new' double-framed 4-4-0s differed from all others of that wheel arrangement in driving wheel diameter. This had started out originally at 5' 0" but became 5' 2" through increased tyre thickness (as with the 0-6-0 Dean Goods). The other 4-4-0s were definitively express passenger locomotives (7' 1" – Armstrongs; 6' 8½" – Atbaras, Badmintons, Cities, Flowers) or mixed traffic machines (5' 8" – Bulldogs, Dukes, Earls [32xx]). It followed that Class 3521 would be focussed on branch and secondary duties.

In tank engine form, Nos. 3521-60 had carried the S2 type boiler (two rings, dome on the front ring and round-topped firebox). The only difference was that the second batch (Nos. 3541-60) retained the slightly shorter fireboxes. As 4-4-0s, the tradition of diversity was sustained through the different types of boiler employed.

Small Boiler Version

Twenty-six locomotives (Nos. 3521-3/ 6/ 7/ 9/ 3530/ 4/ 5/ 7-9/ 3541-6/ 9/ 3550/ 2-5/ 7/ 3560) re-entered service as 4-4-0s carrying S2 boilers fitted with extended smokeboxes. No. 3544 was an exception as it ran for a period with the old-fashioned short smokebox. The conversions were completed:

1899		1900	
January	3553	January	3546
March	3539, 3542	February	3527, 3554
May	3522, 3537	April	3529, 3355
August	3521, 3535, 3549	May	3544
September	3543, 3560	June	3526
October	3523, 3541	July	3550
November	3530, 3557	August	3545
December	3534	October	3538, 3552

Above: This view dated circa 1901 shows No. 3538 at Cardiff. The re-railing jack on the running plate probably saw less use than in earlier years. The tender, which appears to be of the 3000-gallon variety, will receive side fenders in replacement of the coal rails within a few years. *Real Photographs.*

Left: Reflecting the secondary nature of the duties ascribed to the newly created 4-4-0s, some ran with 2000-gallon tenders as with No. 3552 at Pontypool Road circa 1905. Sixty of these tenders were built in four Lots between May 1895 and October 1899, initially for service with Duke Class 4-4-0s in the West Country where short turntables remained in use. They were also associated with the 'Stella' class 2-4-0s and the Dean Goods. No. 3552 engine carried an S2 boiler as seen here until May 1909 when it was replaced by a Type B4. Given the exposed nature of the footplate, the extra weather protection against Welsh rain seems a wise move.

The leading dimensions of tender locomotives originally built under Lot 73 (between Nos 3521 and 3539):

Boiler Barrel	
- length	10' 3"
- outside diameter	4' 3"
- pitch	7' 3"
Firebox - inside length	5' 2"
Heating Surfaces [sq ft]	
- tubes	1070
- firebox	125
Grate [sq ft]	18.9
Boiler Pressure [lb/ sq in]	160
Cylinders	17" x 24"
Tractive Effort [lb - 85%]	15,214
Wheel Diameters	
- bogie	2' 8"
- driving	5' 2"
Wheelbase	5' 6" + 7' 0" + 8' 6"
Weight [tons]	
- total	41.2
- adhesive	27.9
- maximum axle loading	14.6

The differing dimensions for those built under Lot 76 (between Nos 3541 and 3560) were:

Firebox - inside length	4' 8"
Grate [sq ft]	17.2
Boiler Pressure [lb/ sq in]	150
Tractive Effort [lb - 85%]	14,260

Compared with the seismic changes then underway with locomotive design at Swindon, improvement of the flawed 0-4-4BTs was a minor consideration but nonetheless desirable as when the conversion process started in January 1899, the subject machines were only 10-12 years old. The S2 boilers had been fitted new to the 0-4-2Ts/ 0-4-2STs on construction and they were retained with the conversions completed up to October 1900 [Nos. 3538 & 3552]. However by 1900, boiler rotation had become prevalent and with the S2 type approaching redundancy through age, replacements often involved exchanges with 0-6-0 Dean Goods. Between May 1901 and February 1911 eleven engines had their S2 boilers replaced by the S4 Type while No. 3550 whose S2 boiler had received a new firebox in 1903 survived in situ until September 1917.

Below: This undated view appears to have been taken in the early 1920s (note that the Toplight to the left and the more modern coach to the right still have lining on their upper panels). No 3543 has a Type B4 Belpaire boiler which was fitted in December 1912 (superheated from July 1915). The cryptic accompanying notes state 'Birmingham Snow Hill ex Stratford' which suggests its engagement on local passenger duties. ***W L Good.***

The barrels of three S2 boilers were sufficiently sound for their refurbishment and attachment of Belpaire fireboxes, thereby acquiring the comparatively rare B2 designation. They thus enjoyed second lives [No. 3543 – November 1910 to December 1912; No. 3544 – December 1908 to April 1912]. There was often an exception to the rule with GWR classes as exemplified by No. 3530 which carried a B2 boiler from December 1914 to withdrawal in March 1922. Thirteen at some stage carried S4 boilers (some more than once) which was effectively an S2 with the dome placed centrally. With the afore-mentioned exception of No 3530, all eventually acquired the B4 boiler (central dome, Belpaire firebox) which was commonly associated with the Dean Goods in the 20th Century. The different dimensions that applied to engines with B4-boilers:

Heating Surfaces [sq ft]	
- tubes	1092
- firebox	101
Grate [sq ft]	15.3
Boiler Pressure [lb/ sq in]	180
Tractive Effort [lb - 85%]	17,120
Weight [tons]	
- total	41.8
- adhesive	28.8
- maximum axle loading	14.8

No. 3522 in a view where both date and location is unknown but is useful in clearly showing the form of the outside frames retained from tank engine days and the deep valance above the bogie.

Large Boiler Version

Below the running plate this version was identical with the small boiler examples but the superstructure was quite different through fitting of the Standard No. 3 boiler, supported by a built-up saddle and pitched at 7' 10". The height of the Belpaire firebox and of the cab added to the impression of being a substantially larger machine although the increase in total weight was less than four tons while the maximum axle loading was about 14 cwt greater. As with the small boiler version, they were classified yellow for route availability.

Top: No. 3524 at Gloucester Central in 1903 carrying the Standard No. 3 parallel boiler that was fitted on conversion from 0-4-4BT in March 1902. The height of the Belpaire firebox gives the impression of significantly greater size than the small boiler version. ***W Beckerlegge.***

The locomotives so modified were:

1900		**1902**	
June	3528	January	3359, 3547
September	3540	March	3524
December	3551	October	3532
1901			
March	3525, 3548		
June	3531, 3533		
August	3556		
October	3536		
December	3558		

That the combined exercise (small and large boiler versions) took 3 years 10 months to complete reflects the project's low priority.

Bottom: No. 3540 in original condition following conversion with coal rail tender and three panelled monogram livery.

The decision to use the Standard No. 3 boiler appears to have been predicated by either there being no more of Type S2 available, or because refurbishment of any remaining was judged economically unjustified. These more modern boilers were all of the early parallel variety (Type D0) which also had limited life expectancy. Dimensional differences compared with the small boiler 4-4-0s:

Boiler Barrel	
- length	10' 3"
- outside diameter	4' 3" & 4' 5"
- pitch	7' 10"
Firebox - inside length	5' 2"
Heating Surfaces [sq ft]	
- tubes	1437
- firebox	125
Grate [sq ft]	21.3
Boiler Pressure [lb/ sq in]	180
Tractive Effort [lb - 85%]	17,120
Weight [tons]	
- total	45
- adhesive	30.6
- maximum axle loading	15.3

Above: Another early view, of the crew posing with No. 3558, which is paired with a coal-rail tender. Note that in this case the reversing rod has been left exposed.

All fourteen large boiler 4-4-0s received the fully coned version of the Standard No. 3 between February 1907 [No. 3525] and February 1914 [No. 3551], and all carried this type at withdrawal. However, two [Nos 3532 & 3547] reverted to Type D0 in 1910; that on No. 3532 was the last parallel Standard No. 3 boiler in service when withdrawn in June 1915. Nos. 3531, 3458 & 3551 had intermediate spells carrying the short cone version (Type D2) which had originally been used by 2-4-2Ts of Class 36xx.

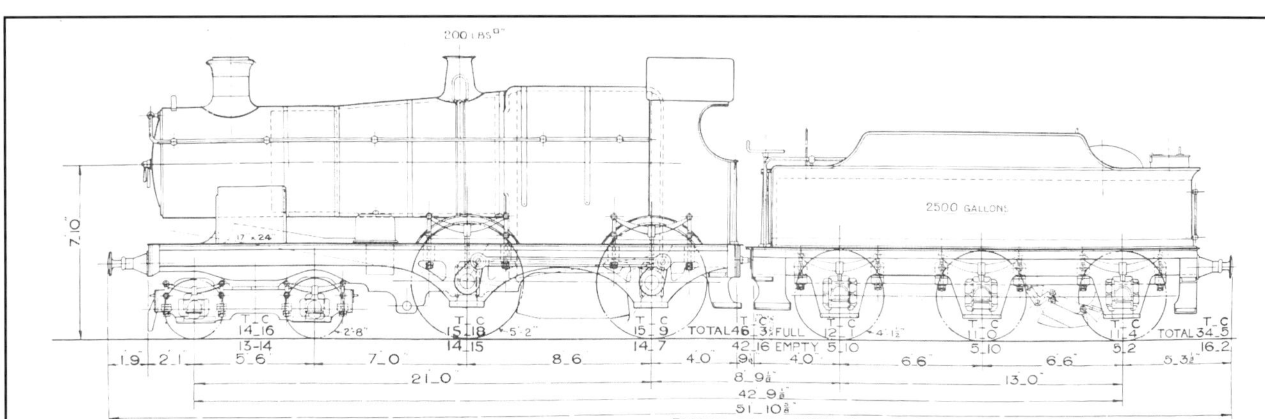

Above: No. 3528 with the superheated, coned D3 Type Standard No. 3 boiler which it acquired in December 1923. The accompanying official Swindon *Diagram A27* gives the principal dimentions. ***GWR Official.***

Superheaters

Superheating was introduced to the class in 1910 but considering the secondary nature of the 4-4-0's duties, this improvement seemed unwarranted. At some stage Churchward gave instructions for the programme to be slowed down. Some superheated boilers were fitted late in their careers, probably through rotation from other classes rather than with any intention to improve performance. There was no reported case of locomotives reverting to saturated condition but some were apparently never superheated.

Small Boiler	
Up to December 1920	3521, 3523, 3529, 3534, 3535, 3537, 3539, 3542-6, 3550, 3552, 3555, 3560
1921 to 1927	3538, 3549, 3553, 3554, 3557
Always Saturated	3522, 3526, 3527, 3530, 3541
Large Boiler	
Up to December 1920	3525, 3528, 3531, 3540, 3547, 3558
1921 to 1927	3524, 3533, 3548, 3551
Always Saturated	3532, 3536, 3556, 3559

Superheaters fitted to the B4 boiler (i.e. Deans Goods type) were probably drawn from the first three batches built between 1911 and 1922 with the following dimensional alternatives:

No. of 5 1/8" Tubes	No. of Elements	Heating Surface (sq ft)
8	48	106
6	48	97
6	36	82

The Type D0 version of the Standard No. 3 boiler remained saturated but with Type D3, the early fittings comprised 12 x 5 1/8" tubes with 72 elements rendering a heating surface of 173 sq ft. Later superheaters used a reduced capacity unit of 6 x 5 1/8" tubes with 36 elements giving a heating surface of 76 sq ft.

Above: It was general practice for second tier locomotives on return from works to be coupled to the next available tender and in this view dated circa 1920, No. 3543 has acquired what appears to be an early 3500-gallon version. *Locomotive Publishing Co.*

Careers as 4-4-0s

The tender engine phase was downbeat compared with the events of the 19th Century. So far as can be determined there was no differentiation of type in relation to assignment of duties. About twenty were first sent to the West Country where some had worked as tank locomotives. They were favoured for service on the Launceston and Falmouth branches, and also employed on assisting duties but had largely been displaced by Duke 4-4-0s and prairie tanks from 1910 onward. They then gravitated to the Worcester division where they were used on longer distance local services radiating from Cheltenham, Gloucester, Hereford and Worcester. By 1919, about half were so engaged with a cluster of about ten based in the Bristol division for similar work. The remainder were distributed between Cardiff, Carmarthen, Didcot, Newton Abbot and Reading.

In the 1920s, Wales became centre of activities. On 26 January 1921, a pair of Cambrian Railways 4-4-0s [Nos. 82 and 95] were damaged beyond repair in the head-on collision at Abermule, Montgomeryshire. Nos. 3521 and 3546 were sold to substitute for these losses and they were allocated but did not carry the numbers of the destroyed engines. In the Grouping process, they returned to the GWR with their identities unchanged on 1 January 1922. More allocations to the ex-Cambrian section followed and these also worked mainly on branch, local and assisting duties. Three of the last six in service (Nos. 3521, 3545, 3554) were located there and the others (Nos. 3529, 3557, 3559) were in the Worcester division. Including their time as tank locomotives, the lowest career mileage was recorded at 626,000 by No. 3541, the first to be withdrawn (in 1914). The last survivor was No. 3557 (withdrawn 1934) which notched up 1,205,000 miles. In view of its comparatively humble duties, an annual average of 26,800 across a career of 45 years was impressive and more than might have been anticipated during the early troubled days.

Left: Back in the fold but still in Wales. No. 3546 was at Machynlleth in 1923, following its short sojourn as a member of the Cambrian Railways' fleet in 1921. This engine had graduated to a B4 boiler in December 1902 but installation of superheating was delayed until November 1915.

Top Left: The '3521 Group' worked during an era of slow camera shutter speeds so views of these locomotives on the move are quite rare. The sparse notes accompanying this undated image state 'GWR 35xx + Duke' and the location is believed to be between Talerddig and Machynlleth. Small boiler Class 3521's became extinct in May 1934, whereas the train engine has the appearance of a named Class 32xx of the short-lived 'Earl' series (dome and straight running plate) of 1936-7. Thus the train engine has to be Duke Class No. 3265 *Tre Pol and Pen* which was withdrawn in December 1929. This engine's boiler was fitted to the frames of Bulldog No. 3365 *Charles Grey Mott* (who had been a director of the company) in early 1930. Re-boilered No. 3265 remained a member of the Duke class as the purpose of the rebuilding had been modernisation while retaining the wider route availability. (By tradition, a rebuilt locomotive was identified by the origin of its frames rather than the boiler). As rebuilt, No. 3265 was effectively the prototype of Class 32xx, the only difference being its use of 3' 2" bogie wheels as opposed to the Earls' 3' 8". The date is therefore sometime between spring 1930 and May 1934, which identifies the pilot locomotive as one of Nos. 3521, 3529, 3545, 3554 or 3557 (the last survivor). The locomotives are forging away with an Up heavily loaded summer service comprising at least nine bogie coaches of varying vintages. The second is a comparatively modern GWR Brake Third whereas the first, and possibly the third have the appearance of Cambrian Railways origin. This is the sort of tantalising photograph that engenders a wish for a better quality camera, if not a time machine. *I A Higgon.*

Top Right: A large-boiler example engaged in less arduous work with an horse box special. The date and location are unrecorded, but the locomotive is No. 3528. *H W Burman.*

Middle: No. 3521 is waiting to depart Aberystwyth on an unknown date with what seems to a local service. It is still coupled to a 2000-gallon tender whose coal rails have since been replaced by side fenders. A B4 boiler was fitted in May 1909 and it was superheated in December 1920, as displayed here. This locomotive was in Cambrian ownership between August and December 1922.

Above: No. 3551 with coned Standard No. 3 boiler, evidently towards the end of its career. This engine was withdrawn in June 1929.

Summary of Allocations

The following is drawn from a handwritten, apparently contemporary, record of class allocations by an unknown author:

Loco No.	January 1901	January 1923	Loco No.	January 1901	January 1923
3521	Frome	Machynlleth	3541	Plymouth	*Withdrawn*
3522	Truro	Worcester	3542	Truro	?
3523	Penzance	Gloucester	3543	Plymouth	Didcot
3524	?	Gloucester	3544	Didcot	?
3525	Didcot	Gloucester	3545	Pontypool Road	Gloucester
3526	Reading	Cheltenham	3546	Cardiff	Machynlleth
3527	Cardiff	?	3547	Malvern Wells	Taunton
3528	Launceston	Reading	3548	Swindon	Landore
3529	Salisbury	?	3549	Cardiff	Didcot
3530	Plymouth	*Withdrawn*	3550	Trowbridge	Bristol
3531	?	Worcester	3551	Plymouth	Landore
3532	Newport	*Withdrawn*	3552	Newport	Yeovil
3533	New Milford	Worcester	3553	Truro	Bristol
3534	Worcester	Carmarthen	3554	Plymouth	Gloucester
3535	Plymouth	?	3555	Plymouth	Gloucester
3536	Abergavenny	Gloucester	3556	Birmingham	Carmarthen
3537	Falmouth	Gloucester	3557	St. Blazey	?
3538	Cardiff	Worcester	3558	Abergavenny	Gloucester
3539	Salisbury	Bristol	3559	Plymouth	Worcester
3540	Didcot	Gloucester	3560	Truro	Didcot

Conclusions

The 20th Century phase of the class's story was largely uneventful, devoted to secondary duties which they handled capably. The span of their working lives as 4-4-0s was not markedly different from those of their better known, larger wheeled brethren, particularly the Atbaras, Cities and Flowers which were culled following arrival of the 4-6-0 Halls. The Bulldogs and Dukes lasted longer, protected by the greater utility bestowed by their 5' 8" driving wheels.

Perhaps the more instructive legacy of Class 3521 was graphically to demonstrate that large tank locomotives were a challenge as so much had to be crammed into the volume prescribed by a single rigid frame, and within the limitations of weight and loading gauge. As noted in the opening paragraphs of Part 1, the company was an enthusiastic champion of the tank locomotive. The traumas that befell the forty pairs of number plates from 3521 to 3560 would have been instructive to William Dean's successor in fresh design initiatives that resulted in the remarkable family of prairie tank locomotives.

Above: End of the road for No. 3521 at Swindon in 1931, with formal withdrawal recorded as September that year. The only evidence of its origins as prototype of this 'group of locomotives' (hardly a 'class') was the distinctive form of the outside frames. The coupled wheel section remained unchanged while working as a narrow gauge 0-4-2T (August 1887-October 1891), then as an 0-4-4BT (from May 1892), before re-emerging as an attractively proportioned 4-4-0 in August 1899.

UPTON IN 1900

BY SEAN BOLAN GRA

ISSUE 9

MODERN TRACTION:
NOCTURNAL 'WESTERNS' IN COLOUR

Successfully capturing railways at night has always provided the photographer with a challenge. The need for a wide camera aperture, coupled with slow shutter speeds, makes even the slightest of subject movement problematical and limits the depth of field within the image. These difficulties were further compounded by the speed limitations of contemporary film stock available to the individuals brave enough to attempt the task in the late steam/early diesel period. Luckily however, when these challenges were faced head on, the results provide us with some of the most atmospheric and engaging of imagery.

Being around trains after dark was an assault on the senses, that seemed to heighten the receptiveness to light, sound and smell. Who can forget the allure of a steam heated carriage, lit by the warm glow from opaque shaded light bulbs. The gentle hum from a diesel engine awaiting the green light to power up and proceed into the darkness, the still air to be filled with that unmistakable exhaust odour before dissipating.

This magical selection of night time photographs is from the collection of Bernard Mills, and features the 'Western' Class diesel-hydraulics during their final decade of operation on the Western Region. The diversity of evening and silent-hours workings shown here is as lamented as the locomotives hauling them, and such scenes are a treasured legacy of those who braved the cold and dark in order to record them.

Many enthusiasts yearn for the days when the railway system was a continuous round the clock running concern. Operating timetables featured passenger as well as freight trains and many stations remained open throughout the night, some with late opening buffet facilities, as well as heated waiting rooms for the replenishment and sanctity of the dedicated. Unfortunately those days are now as much consigned to history as the vehicles that operated them. The very thought of remaining out on the network, whilst maximising the freedom of a 'Railrover' ticket, nought but a distant memory.

Above: The 4.45pm Paddington to Bristol Temple Meads service (1B95) has reached journey's end behind (D)1009 *Western Invader* on Sunday 8 February 1976. The suggested warmth from within Mk.1 Tourist Second Open (TSO) No. W4773, undoubtedly contrasts with the evening temperature being experienced out on the platform.

Above: Exhaust erupts from the two Maybach MD655 power units of a numberplate-less D1009 *Western Invader*, as it departs a deserted Plymouth with the 1B94 6.0pm Penzance-Bristol Temple Meads on 16 October 1976. The long hot summer all but forgotten.

Below: Tripod equipped photographers can be seen capturing D1010 *Western Campaigner* at Yeovil Junction on the evening of 27 November 1976, whilst working the 1Z52 British Rail (Day Excursion) from Plymouth to Birmingham and return via Salisbury.

Above: D1013 *Western Ranger* awaits departure from Fishguard Harbour on the morning of Saturday 25 October 1975, with the Wirral Railway Circle 'Pembroke Coast Express'. The tour ran as 1Z76 throughout (despite the incorrectly displayed headcode).

Below: Exeter St. David's Platform 5 hosts D1013 *Western Ranger* with a short four-coach train. Whilst the identity of the working or date are not recorded, it is after May 1976, when the locomotive received the distinctive red backed name and numberplates.

The famous 1870s vintage trainshed at Bristol Temple Meads, provides a stunning backdrop to D1031 *Western Rifleman* waiting to depart with a late evening Paddington service on 13 July 1970.

D1036 *Western Emperor* waits to depart Plymouth with the 1B19 5.30pm Paddington-Penzance on 20 November 1975. Formed of mainly Mk.2 vehicles, the first pair being Tourist Second Opens (TSO).

D1037

One can almost hear the gentle hum from the engines of (D)1037 *Western Empress*, as it awaits the green light at the east end of Plymouth station on the evening of Tuesday 22 October 1974. The train is the 1A35 7.35pm Penzance-Paddington Travelling Post Office (TPO) working, that had stopped to collect mail and passengers for the late night trip to the capital. The scene simply oozes atmosphere, from the whisp of exhaust fumes and steam escaping, to the warm glow of the light-bulbs in the rear cab and engine compartment. Note also the distant floodlit Mutley Methodist Church and the oil-fuelled tail lamp sitting at the base of the concrete platform light stanchion, waiting to be attached to its next coach rear lamp iron. All long-lost features contributing to the scene.

Above: In the early hours of 7 February 1975, we find D1040 *Western Queen* at the stops on Platform 8 at Paddington. The locomotive had just arrived with the 1A49 sleeper service from Penzance, which departed Cornwall at 10.10pm the previous evening.

Below: The same train as pictured above, is seen at Plymouth some two years later on 19 February 1977. The time is recorded as 0.20am, and a very neglected looking D1041 *Western Prince* performs the duty. A Mk.2 Brake Corridor First (BFK) leads the formation.

Above: D1041 *Western Prince* is seen again, this time on 3 February 1977 at Penzance. The train is the 1E61 Penzance-Sheffield parcels, due to depart at 6.0pm. One of the final five class-members in their last month of service, D1041 was to survive into preservation.

Below: Another locomotive to survive to the bitter end was D1048 *Western Lady*, which is captured at Plymouth on 1 January 1977 working the 6B09 Acton-St Erth milk empties. The eight 6-wheel tankers would have been light work for the 2,700 hp Type 4.

FOOTPLATE RECOLLECTIONS OF A LUCKY MAN

BRIAN WHEELER

In Issue 7 of *Western Times* (Summer 2023) we recounted the recollections of Brian Wheeler and his time as an apprentice and then fitter at Reading, Brian's enthusiasm also creating a number of opportunities for him outside the normal apprentice scheme. As referred to in the first article, at the successful end of his apprenticeship Brian applied for and was successfully appointed on to a supervisory course starting on 19 March 1962. This was for 63-weeks duration and although he was now notionally based at Paddington on-site experience was a considerable part of the course.

He was now issued with a footplate pass and recalls several footplate trips from that time. It is the story of those trips and the notes made at the time that now form the basis for this second instalment of recollections.

Brian's records indicate he made a total of 43 footplate trips between 18 May and 26 October 1962 travelling from Paddington north to Birmingham, Wolverhampton and Shrewsbury, west to Bristol, Weston Super Mare and Cardiff, and south to Eastleigh, Southampton and Bournemouth. Most were by steam although Hymek, Warship, Western and even a lowly diesel unit were also represented. These then are the recollections of a man in a privileged position in the twighlight years of steam on the Western Region:

18 May 1962 - 2.45pm Wolverhampton Low Level to Paddington. No. 6021 *King Richard II* (81A); ten coaches, 358 tons. Driver Gardner (OOC). Departed on time but 7 late upon arrival at Banbury – reason not stated. Maximum of 92 mph at Gerards Cross with cut off at 18% and second valve of regulator. Engine riding and steaming well. Signals at Denham, 5 minute late arrival at Paddington.

With an admiring audience opposite on Platform 1, 'King' Class 4-6-0 No. 6021 *King Richard II* (81A) awaits departure from Paddington on 15 April 1961. ***JLS.***

12 July 1962 - 12.10pm Paddington to Birkenhead (Woodside). No. 6028 *King George VI*. 12 coaches, 411 tons. Driver Dyer (Stafford Road). Footplate as far as Birmingham Snow Hill; only notes that engine rode well with boiler pressure maintained at 230-235 psi. 58mph noted at Princes Risborough.

13 July 1962 - 5.30pm Oxford to Paddington. No. 5065 *Newport Castle* (81A); six coaches, 180 tons. Driver Sneade (Oxford). Departed on time but p/way check at Didcot. Maximum of 88mph near Reading, eased to 75 mph at Twyford. Further signal check at Acton. Arrival at Paddington 3 minutes early at 6.27pm. Noted regulator was stiff, engine riding and steaming well, worked in 15% cut off on second valve.

No. 5065 *Newport Castle* (81A) is captured passing Iver with an Up working on 1 July 1962.

14 July 1962 - 9.45am Paddington - Weston Super Mare (as far as Bristol). No. 7006 *Lydford Castle* (81A), 13 coaches, 435 tons. Driver Ward (OOC). Despite checks at Reading and Didcot, 2 minutes early at Swindon where water was taken. One minute late departure from Swindon, maximum 80mph at Dauntsey where engine eased. Signal checks at Bathhampton and MP 115. Arrival at Temple Meads half minute late. Engine riding and steaming fair, worked at 18% cut off on first valve.

No. 7006 *Lydford Castle* on the down main in Sonning Cutting, 24 July 1950. ***R E Vincent.***

18 July 1962 - Oxford to Eastleigh freight. Churchward 2-6-0 No. 6309 (81E), 39 wagons – no train weight given – 15 vacuum fitted. Driver Potter (Didcot). Departure from Oxford North yard at 3.40pm, one conflat added at Hinksey yard. Stopped for water at Newbury East Junction, passed up freight on single line at Burghclere. Arrived Eastleigh down yard 7.03pm. Engine to shed by SR men. After a break the return trip was with tank cars behind a 9F working a Fawley to Bromford Bridge via Newbury. Brian fired part of the way commenting, 'I did not like the 9F and would have much preferred a 28'. Brian left the train at Didcot and where the Didcot men were also relieved.

19 July 1962 - 12.7pm Gloucester to Paddington. No. 7034 *Ince Castle* (85B), five coaches, 168 tons to Swindon. Driver Beard (Gloucester) Departed Gloucester on time, p/way check at Tuffley Junction, water taken at Kemble. One minute late arrival at Swindon. Six coaches added at Swindon (ex 11.40am from Weston Super Mare), train now 11 coaches, 379 tons. Departure from Swindon on time, 63mph at Didcot but engine eased owing to priming badly otherwise riding and steaming well. Maximum 65mph at Reading. Signal checks at Maidenhead and Acton. Worked at 18% cut off on first valve from Swindon. One minute late arrival at Paddington.

A grimy No. 7034, devoid of its *Ince Castle* nameplates, stands alongside sister engine No. 7029 *Clun Castle* at Gloucester Horton Road shed on 19 February 1965.

20 July 1962 - On this day Brian took what was perhaps a trip best described as 'eventful'. This was on the train engine of the down 'Pines Express' from Bath Green Park to Bournemouth West. The load was 12 coaches, 430 tons, behind No 73051 (82F) in charge of Driver Brewer from Branksome. Assistance was provided by No. 73049 (82F) south from Bath to Evercreech Junction. Because of the length of the train Brian had to walk along the footboards to first reach the engine whereupon he called up to the driver that he had a footplate pass issued from Paddington. The response was best described as 'frosty', for as he was quick to discover from Bert Brewer, Brian represented the dreaded Western Region and who appeared intent on running down the S&D. Departure from Bath Green Park at 4.9pm was on time with both engines worked at 60% cut off and full regulator from Bath through the Devonshire and Combe Down tunnels. Both engines being worked hard with cut offs varying from 70% to 25% for long periods. It was only by Midford that this came down to 55% and at Binegar where it was eased further to 50%. Cut off was increased later to 65-70% upon departing from Poole and remained in this position all the way to the top of Parkstone Bank. Brian recounts that the locomotive steamed well, although the quality of the coal was indifferent. The mechanical condition of the engine was fair whilst the external condition was poor '…for working a train of this class.' A verbal recollection not included on his contemporaneous record was that the atmosphere on the footplate was also 'hot' to say the least. Bert Brewer not impressed that a man from 'the dreaded Western Region' was on board. In the end a mutual love and respect for the steam engine shone through and they parted the best of friends.

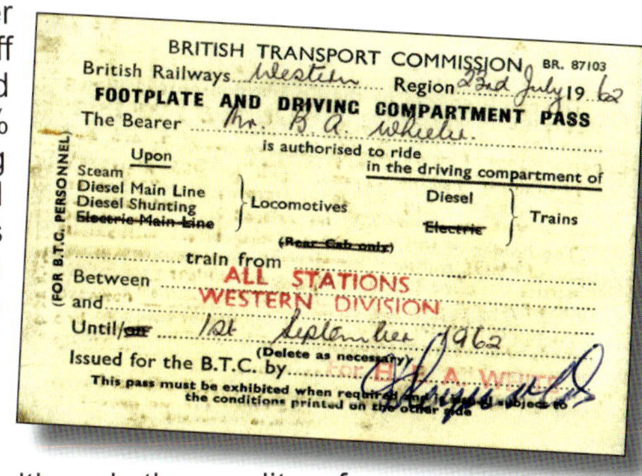

BRITISH TRANSPORT COMMISSION BR. 87103
British Railways Western Region 23rd July 1962
FOOTPLATE AND DRIVING COMPARTMENT PASS
The Bearer Mr. B. A. Wheeler
is authorised to ride
Upon Steam / Diesel Main Line / Diesel Shunting / ~~Electric Main Line~~ Locomotives
in the driving compartment of Diesel / ~~Electric~~ Trains
~~(Rear Cab only)~~
……… train from
Between ALL STATIONS
and WESTERN DIVISION
Until/~~on~~ 1st September 1962
(Delete as necessary)
Issued for the B.T.C. by For B. A. WHITE
This pass must be exhibited when required and is issued subject to the conditions printed on the other side
(FOR B.T.C. PERSONNEL)

25 August 1962 - 8.45am Paddington to Weston Super Mare. No. 6018 *King Henry VI* (81A). 13 coaches, 432 tons. Driver Battams (OOC). Departure on time with a maximum of 70mph before Reading. Water taken on the troughs at Goring and again during the stop at Swindon. Three signal checks en-route and a 20mph restriction for relaying on Dauntsey bank. Two minute late arrival at destination (12.02pm). Engine steaming well but only riding fair. 18% cut off on first valve.

Upon arrival at Weston, Brian quickly swapped footplates on to No. 45725 *Repulse*, on what the first leg of the through train from Weston to Sheffield which he rode as far as Bristol. The load was 10 coaches, 327 tons with Driver Glasspool of Barrow Road in charge. Departure was on time at 12.15pm but signals brought the train to stand just four minutes later (this was due to the late running of the 12.08pm ex Ilfracombe service also from Weston bound for Manchester). A maximum of 64mph was achieved past Flax Bourton with the train 9 minutes late into Bristol. Brian's comments on this LMR machine are not particularly complimentary. 'Engine steamed well but riding rough…..not fit for passenger work due to excessive play in trailing boxes etc. Engine worked in 20% cut off with second valve when recovering from 25-30mph to 60mph.'

There was a third trip on the same day enabling Brian to return to Reading on what was the 1.15pm Bristol Temple Meads to Paddington. This was behind No. 5943 *Elmdon Hall* (82C). The load was 12 coaches, 374 tons. Driver Battans was again in charge. There was a one minute late departure from Bristol after which the engine performed badly. Running into Bath the driver commented upon a man wearing a raincoat and trilby waiting at the end of the platform and who joined us on the footplate. This was Kenneth Leach and who was immediately invited to drive the train to the next stop at Chippenham where he lived. Brian was also invited to fire over the same stretch of line and which included the climb through Box. The engine was not steaming well but we plugged away. Upon arrival at Chippenham, Kenneth took a photograph of the three of us which is reproduced below. At this point the driver said Brian could now drive the engine as far as Wootton Bassett which he did; a most enjoyable experience. Unfortunately, though the steaming capability of No. 5943 was such that even allowing for a signal check at Wootton Bassett we were 19 late by the time it reached Reading. Brian's commented that apart from the steaming the ride was fair. A maximum speed of only 55mph was achieved and this over the troughs at Goring – notwithstanding half a tank of water having been taken only a short time before at the Didcot stop. The late running was put down to the condition of No. 5943. To add to the woes upon arrival at Reading it was necessary to have one of the coaches examined for a suspected hot axle box.

The 25 August 1962 photograph of No. 5943 *Elmdon Hall* paused at Chippenham. Brian is seen in the middle. *Kenneth Leach.*

3 September 1962 - Brian again made more than one run. The first was the afternoon, 1.07pm, from Reading to Oxford. A train that had commenced at Margate and would eventually reach Wolverhampton. Reading based No. 5076 *Gladiator* (81D) was in charge having taken over from SR 'Schools' 4-4-0 No. 30901 *Winchester* on a load of 12 coaches, 409 tons. Driver Stephenson (Oxford) was on the footplate. Arrival from the Southern had been scheduled for 12.57pm but the train was 33 minutes late and did not appear until 1.30pm. Reading was quick to exchange engines and No. 5076 was away at 1.39pm after which it was an easy run to Oxford with a maximum of 65mph at Cholsey although arrival at Oxford was still 30 minutes late. Brian notes the engine both steamed and rode well and was in first class mechanical and external condition.

No. 5076 *Gladiator* passing Reading shed bound for Paddington on 7 July 1962. *JJ1641.*

The return run from Oxford was the complete opposite compared with the condition of No. 5076. This was with No. 7009 *Athelney Castle* on the late running 2.39pm Oxford-Paddington. Indeed, departure was not until 3.10pm reported as due to the failure of the engine on the proceeding train. No. 7009 had nine coaches weighing 312 tons, the service having originated from Hereford. Driver Kite from Worcester was on the footplate. Having got away speed was worked up to 65mph at Goring to take water and then 75mph at Scours Lane on the western outskirts of Reading. 78/79mph was achieved at Slough with the engine shut off at Acton and allowed to drift the final four miles to the terminus. A net time of 1 hour 23 minutes was three minutes inside the original schedule. As such it might appear as if this was a satisfactory run but not so for Brian writes, 'Engine riding rough 'AFLOC' fitted. Coal a mixture of slack and nuts. Cab etc shaking so badly that the ATC bell could not be heard – firehole flap completely adrift after Reading.'

The second of the British Railways constructed 'Castle' Class locomotives (May 1948), No. 7009 *Athelney Castle* at Hinksey South on 25 August 1962. *R C Riley (RCR 16757).*

4 September 1962 - Another 'out and back' run was made, starting with the 7.10am Didcot to Paddington which Brian joined at Reading. No. 6978 *Haroldstone Hall* was at the head of 12 coaches weighing 368 tons with Driver White also from Didcot. Departure from Reading was on time although a signal check at Old Oak Common East of just over a minute caused a similar late arrival into Paddington – there being no opportunity to make up time over the remaining short distance. A maximum of 62mph was recorded at Maidenhead. Again, this was with an engine in not ideal condition, reported as steaming badly due to blocked tubes, the riding rough and the ATC not in working order.

Brian again rode with Driver White for his outward journey from Paddington; the 9.15am Paddington to Hereford which he reported on as far as Worcester. Driver White took the train of nine coaches, 317 tons, to Oxford after which Driver Kimble from Worcester was in charge. When asked Brian regrets not having recorded the names of many of the firemen he rode with. Departure at 9.15am was on time but again there was a signal check at Old Oak Common which cost two minutes. Despite working the engine up to 70mph beyond Twyford, speed continued to be lost and the train was just over three minutes late into Reading. Brian was later to be a bit critical of Driver White commenting in his notes, '….. right time arrival at Reading would have been possible if the engine had been worked harder during early stage of run.' Between Acton and Reading 18% cut off had been used on the first valve of the regulator. The section of line from Reading to Oxford was more easily timed (recovery time perhaps?) and Oxford was reached 3 minutes early at 10.30am. Departure was one minute late after which things proceeded smoothly as far as Blockley when the train ran over detonators placed on the track. No emergency situation was reported so it was likely these had been placed to protect an earlier situation and simply not collected afterwards. A further delay was due to a p/way check just prior to Pershore culminating in a 5 minute late arrival. On the plus side the engine both steamed and rode well.

Having reached Hereford Brian needed to return home and this was achieved courtesy of a Hereford – Worcester train, changing there to a Worcester – Paddington working. The 19 miles from Hereford to Worcester was behind No. 7928 *Wolf Hall* (85A) which left at 4.05pm on just four coaches weighing 134 tons and with Driver Pritchard from Hereford on the regulator. Just 4 miles out however they were brought to a dead stand for signals at Shelwick for I minute caused by a freight coming off the line from Shrewsbury. Resuming, No. 7928 was worked up to 72mph by Ashperton which enabled time to be regained and meaning its scheduled stops at Ledbury, Great Malvern, Malvern Link, and Foregate Street were all to time with a final arrival at Worcester Shrub Hill just less than 1 minute early. It was an excellent journey, albeit short with the engine riding and steaming well.

Just 10 minutes after arriving Brian was off again, taking the 5.10pm from Worcester to Paddington behind No. 7011 *Banbury Castle* (85A) in charge of eight coaches weighting 276 tons. Driver Perry was in charge. There was a p/way check at Fladbury otherwise this journey was without incident with a right time arrival at Reading of 7.15pm. Water had been taken twice, at Charlbury and then again at Goring. The on-time arrival assisted by being a minute or so early at the various passing points south of Oxford but which advantage was then lost by signals at Scours Lane on the outskirts of Reading with brought the speed down to just 10mph. Brian noted the riding was fair but which he commented was '…not up to the usual Worcester standard'. The engine steaming well on a mixture of slack, nuts and lumps.

Locally allocated No. 7928 *Wolf Hall* stands at Worcester Shrub Hill in the summer of 1963. ***Neville Stead (NS 201159).***

6 September 1962 - Two days later it was a freight turn with BR Standard No. 73012 (82C) initially on 58 non-fitted wagons from Didcot to Westbury routed via Chippenham and Thingley Junction. No train weight is given. Driver Jones from Didcot was in charge. The engine was started from Didcot North with 65% cut off and first valve but not for long as soon after starting they were brought to a stand for signals at Foxhall Junction. Restarting the train continued west and by Wantage was running at 30% cut off increased to 35% and second valve at Challow. Brian then has the entry '35 – Swindon Down yard', which we may take to mean this was the reduced load taken from Swindon with no further entries other than an arrival time at Westbury Down Yard of 3.40pm. Brian recalls he was driving for part of the route and was in charge of the stop at Swindon, he admits 'It was not one of my best performances with the steam brake.'

Brian was basically free to choose his trips and experiences, and as will be gathered, he initially rode on steam but also knew he would be questioned if he did not include some experiences on the more modern types of traction. Consequently, his next outing was with diesel on **7 September 1962** but only on a single unit car, No. W55035. Operating the service between Didcot and Newbury in what was also destined to be the final week of passenger operation over this line (passenger services were withdrawn just three days later). Departure from Didcot with Driver Haycroft was at 10.50am, calling at all stations with the exception of Churn and Pinewood Halt, to arrive at Newbury on time at 11.30. No observations were made although it would be fair to say traffic was probably very light – an oft quoted phrase being that a lack of passengers has the bizarre advantage of leading to good timekeeping.

26 September 1962 - Was to see another trip over the DNS line starting light from Didcot to Oxford with Mogul No. 6309 (81E) and driver Dodd of Didcot. At Oxford they turned and then backed on to a train of 35 wagons for Eastleigh. Leaving at 3.40pm the journey south took until 6.25pm and they passed an Up freight service at Burghclere. Apart from commenting the engine was in good mechanical condition and steamed well, no other mention is made – nor of how he returned to Reading.

28 September 1962 - Another trip without comment on the locomotive was made on No. 5987 *Brocket Hall* (81E) working parcels from Didcot to Paddington. Initially the load is stated as having been five vans equal to 121 tons but at Reading that changed although still with five vans to 171 tons. Driver Wren from Didcot was on the engine. The train left Didcot 5 minutes late at 1.05pm and was still 3 late at Reading. Here 18 minutes were allowed and this was cut to 15, despite the load difference that now applied. Stops were made at Twyford, Maidenhead, Slough, West Drayton, Hayes, Southall and Acton with arrival at Paddington 7 minutes early at 4.41pm. Brian noted the train was swapped from relief to main at Cholsey and then back to the relief lines again at Scours Lane. There was also a p/way check at Pangbourne and the use of the Up loop at Maidenhead. His return from Paddington was his final recorded run on a 'King', No. 6011 *King James I* (81A). This was a Didcot service of eight coaches weighing 256 tons. Driver Wren also took charge back to his home depot. The start from Paddington was two minutes late and 74mph had been reached at West Drayton before a series of signal checks at Langley and then Burnham reduced progress. Arrival at Reading was just over 2 minutes late with Brian commenting the engine rode and steamed well and was 'AFLOC' fitted. 'A right time arrival was impossible due to the late departure coupled with the pair of signal checks.'

'King' Class No. 6011 *King James I* with a down Wolverhampton working passing Old Oak Common East on 12 October 1957. ***R C Riley (RCR 11324).***

2 October 1962 - A case of from the 'sublime to the ridiculous', with a pannier tank journey – no one could say Brian was not taking every opportunity to gain experience. This was on the 7.15am freight from Didcot to Hungerford via Compton behind No. 8720 (81E) and Driver Goodall. This journey was also the typical pick-up freight for the train shunted at Compton, Hampstead Norris, Hermitage (Government cold store), Kintbury and Hungerford on the outward journey – having also stopped for water at Newbury West in the interim. The service left Hungerford for the return with 16 wagons again calling at Newbury for water. Once on the branch there was further shunting at Hermitage, Compton and this time Upton & Blewbury before returning to shed at Didcot for 2.30pm. No other comments are made nor the number of wagons perhaps deposited or collected at each point.

Away from the glamour perhaps, Collett 57xx 0-6-0PT No. 8720 at Didcot shed on 25 July 1964. ***Alec Swain.***

4 October 1962 - For his next trip Brian travelled on an engine that was ex works from Swindon on a running-in turn. This was No. 6966 *Witchingham Hall* (81A) with 30 wagons and a turn he had previously experienced; Didcot to Westbury via Thingley Junction. Driver Smith from Swindon was initially in charge with a crew change to Westbury men at Swindon. This was stated to be the reason for a delay of almost 90 minutes in Swindon Down yard. The total journey time was from 11.40am to 3.35pm with no other comments made.

Nameplate of No. 6966, whilst carrying BR mixed traffic lined black livery, taken on 12 January 1954.

Also on 4 October 1962, Brian made the first of his runs on a main line diesel, with 'Warship' No. D863 *Warrior*. The engine was on a Bristol Temple Meads to Paddington working, eight coaches at 277 tons with Driver Criddle of Bath Road (Presumably for this and his subsequent diesel trips there would have been a second man on board but this is not mentioned). The start from Bristol was 6 minutes down at 5.40pm with the train continuing to lose time so that it was 8 minutes down by the Chippenham stop. Some recovery was then made but this was hampered by signals at Rushey Platt which brought the speed down to 22mph. Things might have improved had the 3 minute timing for the stop at Swindon been adhered to but instead this was extended to 5 minutes and the eventual arrival time at Reading (where Brian left the cab) just after 7.01pm, 6 minutes down. The minimum speed on Dauntsey bank was 64mph working on notch 6 after which it was mostly notch 3 or 4 which achieved the maximum permitted 80mph for the class maintained from Wantage through to Didcot East Junction.

9 October 1962 - Five days later Brian recorded his only run on the footplate of a Hawksworth 'County', No. 1015 *County of Gloucester* (81E). This was with the 7.10am Didcot to Paddington loaded to 12 coaches equal to 368tons, which Brian joined at Reading. On this occasion we do have names for the driver and fireman, respectively Messrs Johns and Fuller from Didcot. The service was 5 late leaving Reading and maintained this degree of running to arrive at Paddington similarly 5 minutes down. For once Brian did not record any speeds only the schedule and passing times but he comments 'Engine riding and steaming badly, average boiler pressure maintained at 170psi. Engine worked at 25% and on second valve – fog patches'.

In better condition than when Brian experienced her, No. 1015 *County of Gloucester*, ex-works at Swindon on 1 April 1962. Six months later Brian reported this engine as riding and steaming badly with pressure maintained at just 170psi. *Alec Swain.*

Arriving at Paddington, Brian now joined No. 7017 *G J Churchward* at the head of the 9.15am Hereford service. The same crew were on board for a train of ten coaches weighing 342 tons. The start was on time with the engine quickly noticed as riding and steaming well (a definite improvement on No. 1015 a short time earlier). Brian was on the train as far as Oxford but noted the maximum speed achieved was just 68mph at Slough. A signal check early on at Westbourne Park meant a late arrival and departure from Reading was followed by a further signal check at Moreton Cutting. The delay at Reading had been due to a freight working crossing their path. Even so the crew and engine worked well, water was taken at 60mph on Goring Troughs and despite the delays they were only 1 minute down at Oxford. The engine had also been worked lightly between 18% and 15% cut off on the first valve.

10 October 1962 - The following morning it was another early start, with the same working from Reading to Paddington and likewise Paddington to Oxford. In charge on the Up run was the same Didcot crew this time with No. 4959 *Purley Hall*, likely the same rake of coaches was also used as the number and weight was also identical. No. 4959 was a better engine as well as steaming well although the riding is described as only fair. Departure from Reading and arrival at Paddington were late due this time to fog working (double block) being in force.

For the Oxford journey it was the turn of No. 5093 *Upton Castle* described by Brian as riding and steaming well. On this journey the departure from Paddington was prompt at 9.15am and despite there being no reported slowing on the way, Reading was reached 2 minutes late. Signal checks then occurred at Moreton Cutting and Radley resulting in a 2 minute late arrival at Oxford. Maximum speed had been just 65mph at Maidenhead with the engine worked on the first valve at 15% cut off, boiler pressure being maintained at 215/200psi.

11 October 1962 - Saw a footplate run from Swindon to Paddington behind No. 5054 *Earl of Ducie* (87A), an engine that found fame two years later in 1964, but was destined to end up cut into tiny pieces and recycled. For this working it was in charge of 11 coaches, weighing 383 tons, with the 9.53am service. Didcot crew Driver Harris and Fireman Cox were working the locomotive. The journey is perhaps best described as leisurely, the engine riding and steaming well and again with boiler pressure maintained at 215/220psi, just below the blowing off point. The maximum speed reported was 68mph at Slough followed by a signal check at Hayes which brought the train to a dead stand. The crew did well then to be less than 2 minutes late arriving.

No. 5054 *Earl of Ducie* passing Moreton Cutting east of Didcot, with a service from Worcester on 1 August 1964. *Brian Wadey.*

12 October 1962 - Brian had some scathing remarks to make about his next experience on No. 4908 *Broome Hall* (81E). Once more this was on the 7.10am ex Didcot with an identical load and Driver Johns. The journey is best summed up in his own words, 'Engine riding rough, steaming badly. Engine not considered fit for this class of train. Late arrival (it was just 2 minutes) due to engine condition'.

The Oxford return was on No. 7018 *Drysllwyn Castle* ('AFLOC' fitted), ten coaches, 343 tons. This time the train was 2 minutes early at Oxford with a maximum speed of 70mph at Kennington Junction. The loco steamed well but rode rough, being fed a diet of slack and nuts. Again, the cut off was identical to other locos on this working, 18% to 15% and first valve on the regulator.

Still wearing its final GW livery, No 4908 *Broome Hall* is working an Up suburban service near Iver on 29 February 1952.

15 October 1962 - Brought another diesel turn, No. D869 Superb with the 9.45am Paddington to Weston Super Mare, which Brian joined at Didcot. Driver Say of OOC was in the cab with eight coaches weighing 256 tons behind. The train was slightly less than 3 minutes late leaving Didcot but had achieved right time by Bristol and an arrival 2 minutes early at destination. A maximum speed of 78mph was recorded at Dauntsey with notch 6 on the controller only used on two occasions, first for a short time at Wantage Road and then again at Thingley Junction.

17 October 1962 - Two days later it was to a new destination, Cardiff, in the cab of 'Hymek' No. D7037 working the 7.55am Paddington to Cardiff, which Brian joined at Reading. The load was just six coaches for 208 tons with Driver Baker of Canton. The weather was not kind, dense fog, meaning the train was already 6 minutes late at Reading. Some time had been made up as they turned on to the South Wales main line at Wootton Bassett but a signal check at Little Somerford cancelled out any advantage gained. After this there was an attempt to regain time with 70mph slightly further along at Badminton but they were brought to a dead stand at Cattybrook Sidings Signal Box for one minute with continuing delays after this meaning arrival in Cardiff was 17 minutes down.

Returning to Reading, Brian was aboard another 'Hymek', this time No. D7038 with the 8.00am service from Neyland. He reports that fog was initially patchy but later cleared, meaning Driver S Jones from Canton was able to keep his 11 coach train of 383 tons to time, notwithstanding a signal check at Swindon West. Departure from Cardiff was prompt at mid-day and arrival in Reading 1 minute early at 2.24pm. A maximum of 78mph had been achieved at Wantage Road.

Beyer Peacock Type 3 'Hymek' Nos. D7038 and D7002 outside Swindon Works 'A' Shop on 24 March 1963. *Alec Swain.*

19 October 1962 - Brian again broke new ground with a return trip on a 'Western' Class diesel-hydraulic between Paddington and Shrewsbury. Outward was with No. D1004 *Western Crusader* and Driver G E Owen of Shrewsbury. The load was 12 coaches at 445 tons; the train the 11.10am 'Cambrian Coast Express'. This was a run about which Brian makes no comment, suffice to say he left on time (11.10am) and arrived on time (2.14pm) despite three out of course delays. The first was a signal check early on at Old Oak Common, then a slowing to 30mph through the platform line at High Wycombe and finally a p/way check between Hatton and Lapworth. A maximum of 88mph was achieved at Fosse Road.

Returning from Shrewsbury was with 'The Inter City', the same locomotive in charge and this time with Driver Bench from Stafford Road. The load was slightly reduced with 11 coaches and 368 tons. Again, it was a faultless performance, a 2 minute late departure at 3.54pm translated into a 4 minute early arrival at Paddington (6.55pm). No delays were reported on the way and the maximum speed was 84mph at Bicester North.

When Brian travelled on green liveried diesel-hydraulic No. D1004 *Western Crusader*, it was barely five months old. *Brian Wadey.*

22 October 1962 - It might be thought that the majority of steam runs were now becoming tasked to engines in poor condition, but Swindon was still at the time overhauling steam and Brian's next turn was with another ex-works engine, No. 6869 *Resolven Grange* for the short journey between Didcot and Swindon. The train was the 7.55am Banbury to Westbury freight loaded to 39 wagons (no weight is given) with Driver Carron from Didcot. The only comment made being the train was put into the down loop at Uffington and remained there for 46 minutes either due to passing traffic or perhaps being unable to be accommodated further down the line. No. 6869 was worked on the first valve at 25% cut off.

23 October 1962 - Saw Brian on the footplate of the engine working the down 'Pines Express', Class 5 No. 73042 (71G) with Driver Guy from Branksome (a former S&D man). This service had now been diverted off its former S&D route to run via Oxford and Basingstoke with Brian joining the footplate at Oxford and travelling as far as Southampton. No. 73042 rode fairly and steaming was average, but was not in good external condition. The schedule was kept (Oxford 2.04pm – Southampton Central 4.00pm). One signal check was reported at Reading West and a maximum of 76 on the falling gradient between Micheldever and Winchester City. Perhaps because of the load, 12 coaches at 434 tons, the working was also hard with cut offs between 18% and 45% and a mixture of first and second valve.

24 October 1962 - A further 'Hymek' run took place from Didcot to Paddington, No. D7042 and Driver Farr of Bath Road in charge of the 8.53am service from Taunton. The load was 12 coaches for 434 tons. This train was also non-stop through Reading which was passed at 74mph. A maximum of 82mph was reached at Burnham after which the engine was eased. Despite a p/way slowing east of Slough and a signal check at Old Oak, the arrival at Paddington was 3 minutes early at 12.25pm.

Brian's penultimate run with steam was the same day, returning from Paddington to Oxford with the 1.15pm Hereford service behind No. 7031 *Cromwell's Castle* (85A). In charge was Driver Ashby from Worcester. The load was nine coaches weighing 314 tons. *Cromwell's Castle* was an example of how steam could and should be, riding and steaming well and in good external condition. Paddington was left on time and Oxford reached 2 minutes early and with just one slight delay at Didcot East which brought the speed down to 15mph. Otherwise a maximum of 75mph was recorded at Slough working at 15% cut off on the second valve.

For the return from Oxford, it was on to the footplate of No. 7027 *Thornbury Castle* (85A) loaded to ten coaches weighting 342 tons. On this occasion the driver's name was not recorded. *Thornbury Castle* rode well but steaming was only fair, externally it was good. Brian noted it was 'AFLOC' fitted with the coal the regular mixture of slack and nuts. The train was three minutes late departing from Oxford at 2.42pm and maintained a similar degree of lateness to Reading. By the time they reached Southall they were 7 minutes down the situation not helped by a signal check at Acton. Arrival at Paddington was just over 7 down with an earlier maximum speed of 68mph at Slough.

26 October 1962 - Brian recorded his final footplate experiences. The first run was from Reading to Paddington, with No. D7021 in charge of seven coaches (219 tons) working the 7.10am from Trowbridge with Driver WIlliams of Westbury. A maximum of 74mph was reached at West Drayton, but the trip was generally hampered by a succession of signal checks, resulting in a 6 minute late arrival in London.

The return was on the 11.45am Paddington to Taunton via Bristol. This had No. D854 *Tiger* as the train engine with No. D7052 assisting. The load was ten coaches at 353 tons, with Driver Bakeman of Swindon. It might have been thought that with some 3,700hp available from the two locomotives, time keeping and if necessary recovery would be easy, but it was not to be, and Brian was to experience probably one of his worst runs. A signal failure at Old Oak cost them 10 minutes with a further slowing for p/way work at Southall bringing speed down to 20mph. Arrival at Swindon was 16 minutes late at 1.25pm.

Brian's original records of these runs together with other paperwork from his railway life are now deposited at the STEAM Museum in Swindon.

OXFORD IN WORLD WAR 2

The recent discovery of a torn and shabby poster dating from WW2 has led to an amount of speculation and not a few questions. Fortunately, enough has survived to provide some answers but similarly others have been raised. Admission to the exhibition would almost certainly have been free.

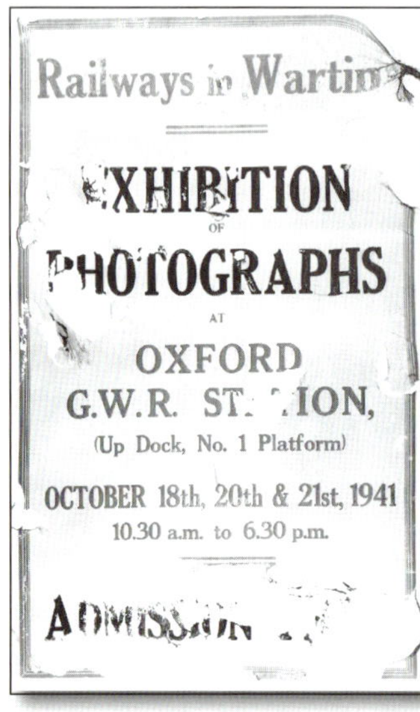

First of all, and perhaps the most obvious, why would the Government want such an exhibition and perhaps even more important, what might it depict? So far as the first question is concerned it has to be morale – official propaganda if you prefer. As regards what it might depict perhaps it would be easier to start by excluding what it certainly would not show; dishevelled passengers, bomb damage; the latter at first almost certainly discarded but in that respect have a thought. Suppose it did show a damaged area of railway and then the restored service, surely this would be a good morale booster indicating 'the railways can cope'. Another possibility might be how the railways are capable of manufacturing and moving vast quantities of war materials, the workshops converted for war production, and women working in the various grades and departments.

We would have been less likely to be shown new works, 'progress' as the GWR itself referred to such endeavours, whilst it will also be noted the theme is 'Railways in Wartime' and not 'The GWR in Wartime'. Most likely then this was national exhibition of images and a mobile one too as the exhibition is indicated as being in the Up Dock of Platform 1, so implying it was located in a railway vehicle.

Oxford was fortunate, as it was never the subject of Luftwaffe bombing in WW2. Folklore has it that Hitler spared Oxford as he intended the city as a major location in his victorious Third Reich. Other supposition has it that an agreement existed between England and Germany not to attack particular historic towns; but this did not stop Cambridge, Bath and Canterbury being targeted. It also seems slightly strange that in the midst of all-out war a 'Gentleman's' agreement somehow might still exist. Whatever, Oxford played its part in many ways, housing evacuees, government departments, and of course increased railway facilities at what was a vital interchange point as well as part of the north-south artery feeding munitions and supplies to the south coast ports.

'Progress' as it was referred to, would probably have been recorded by the official GWR photographer – such material sadly now buried in the bowels of the NRM with access difficult. However, what we do have is access to a small collection of images taken, we think, by R H G Simpson around Kennington, just south of Oxford during WW2. We say 'we think' for as some will know, Mr Simpson was a supplier of railway photographs for several decades post WW2. Certainly in later years he also sub-contracted others to record the railway scene and whose images were subsequently assimilated into the general R H G Simpson archive. Might these 1940s images also then be by another hand?

Considering the general fear of invasion that existed in the early years of the war and the ongoing threat of spies and fifth columnists that similarly existed, the photographs are a remarkable record. They are undoubtedly not the best quality technically, but then they are not intended to show a detailed record, but instead to present an insight into the trains and general conditions pervading at the time. For that value alone, they are worthy of preserving and sharing.

Right: The majority of the R H G Simpson images of the time date from 1944/5. This is one of the exceptions with the unusual combination of No. 6873 *Caradoc Grange* and 0-6-2T No. 6673 heading north past Kennington Junction Signal Box on 31 August 1941.

Much new work in the form of additional sidings and loops to deal with the increasing wartime traffic took place in this area from about 1942 onwards, in particular north of Kennington and in the vicinity of Hinksey. *R H G Simpson.*

Above: Government stores (which could literally be anything and everything) and personnel trains specifically for the military, were two types of working that had to be accommodated as a priority over ordinary traffic. The latter were liable to cancellation at short notice whilst the former would receive precedence. In consequence it is easy to see how ordinary passenger traffic became difficult, delayed and crowded. This is an example of a 'government stores' working, No. 5036 *Lyonshall Castle* unusually the motive power for a long train of open wagons between Hinksey and Kennington on 15 April 1945. *R H G Simpson.*

Below: Many of the Simpson images were taken at, near to, or from the signal box at Kennington Junction (proving the point it is not what you know, but who you know) such as here with No. 6906 *Chicheley Hall* in charge of a down military working; two LNER type coaches and flat wagons carrying tanks. It is possible this was running unfitted (another similar view from the collection, unusable due to blurring, shows a 53xx with a brake van at the rear) and yet it will be noted it is also operating under 'Class A' headlamps. Both images indicate the necessary priority given to such workings. 27 April 1944. *R H G Simpson.*

Ambulance Trains were another priority service. One such working is recorded behind LNER B12 4-6-0 No. 8557, being piloted by tender first Churchward 2-6-0 No. 5303, again with 'Class A' lamps on 25 June 1944. The train is coming off the Thame branch to join the main line northbound. A number of Ambulance Trains operating in the UK were formed of Westinghouse braked stock, hence the B12, and were operated throughout by two eastern crews and a fitter billeted on the train itself. The ex-GER design was also favoured for such trains as their short coupled wheelbase (14') and maximum axle loading (17 tons) gave them wide route availability. In May 1944 the GWR issued a booklet giving point to point times and list of hospitals and ambulance railheads. All the home Ambulance Trains were made up of 14 vehicles, the detailed formation depended upon whether they were classified as a 'Home Ambulance Train' or a 'Casualty Evacuation Train'. *R H G Simpson.*

Left: A similar working this time with No. 6827 *Llanfrechfa Grange* piloting another B12, No. 8525. The train is on the main line approaching Kennington from the north. The formation of such workings involved a stores vehicle, a restaurant car, ward cars and accommodation for the medical personnel and train crew. Again, the 'Class A' headlights apply. Understandably engine cleaning was not a priority at the time. Aside from access to take his photographs, we have to wonder where Mr Simpson obtained film. In addition, he would certainly have been noticed by the trackside recording information (nearly all the views have engine details and a date) so it is a wonder he was not at least questioned by the authorities. *R H G Simpson.*

Below: What appears to be a new Ambulance Train working as an unfitted service. The engine is No. 5998 *Trevor Hall* photographed on 14 March 1944, with a service that may well be bound for France – in which case the Westinghouse brake only would be fitted. Conversion of carriage stock into Ambulance Trains took place at several works and appears to have involved either GWR or LMS stock. This particular formation is probably on the way south ready to be shipped across to Europe as soon as the forthcoming invasion was successful, and a foothold had been achieved. *R H G Simpson.*

Above: A final Ambulance working is seen comprised of conventional vacuum brake stock behind No. 5964 *Wolseley Hall* near Oxford on 16 May 1944. Whilst the official railway photographers and similarly various press photographers of the period recorded wartime damage, new works, passenger evacuations, newsworthy items etc, there are very few records of regular day to day workings during the period. Mr Simpson's views we accept are certainly not perfect, but they are unique, covering a period when of the tens of thousands of special trains run, probably less than a handful were photographed. *R H G Simpson.*

Below: Wartime operations did not always go smoothly either, as this is a derailment at Kennington on 28 March 1944 testifies. We also know there were accidents at Sandford and Appleford in WW2 (the late Harold Gasson speaks of one in particular in his memoires), but otherwise there is no further information pertaining to this particular incident. *R H G Simpson.*

Above: We conclude with the unusual sight of No. 5010 *Restormel Castle* with two former 'Dean Goods' 0-6-0s in tow, WD Nos. 70094 and 70095, heading south at Kennington early in 1945 and probably bound for Swindon. Formerly GWR Nos. 2399 and 2470 (WD Nos. 94 & 95 respectively) these were two of a batch of eight 'Dean Goods' lent to the War Department by the GWR in the autumn of 1940; part of a batch of engines from the class subsequently sold to the government in December that year. By the end of 1940, Nos. 94 and 95 were in Kent in conjunction with the rail-mounted heavy artillery guns at Martin Mill near Dover. No 94 later received an extended roof, the only significant modification applied to the batch of eight while in UK service; it was also christened Monty. The 'Dean Goods' under UK Government control had their original WD numbers increased by 70000 in 1944. (Although far from pristine here, it cannot be confirmed whether these were the pair in Kent that were kept in sparkling condition with red coupling rods. On discovering that the domes were brass, the sappers scraped off all the paint and polished them to pre-Grouping glory before such antics were deemed un-military.) ***R H G Simpson.***

As a footnote to the two War Department 'Dean Goods' locomotives pictured above, their known deployments and dates are:

No. 94 (ex-2399): Faslane Scotland [Nov-1941]; Military port, Cairnryan [1941]; Bicester Central Ordnance Depot [Aug 1942]; collided with No. 95 [Dec 1944] and sent to Swindon for repairs [Feb 1945]; SRD works [Sep 1946]; Shropshire & Montgomery Railway [Dec 1946]; scrapped 1948.

No. 95 (ex-2470): Military port, Cairnryan [1941]; Kent [1943] Bicester Central Ordnance Depot [Aug 1942]; deployments during 1944 - East Kent/ Longmoor Military railways and then Bicester Central Ordnance Depot [Aug 1942]; collided with No. 94 [Dec 1944] and sent to Swindon for repairs [Feb 1945]; Shropshire & Montgomery Railway [Dec 1946]; scrapped 1948. (Whilst at Bicester, it was not uncommon to see a Dean Goods in WD livery working a leave train from Bicester to Oxford.)

N.B. No. 2470 was also on loan to the government from early 1917 and served in France. It returned to the GWR in June 1919.

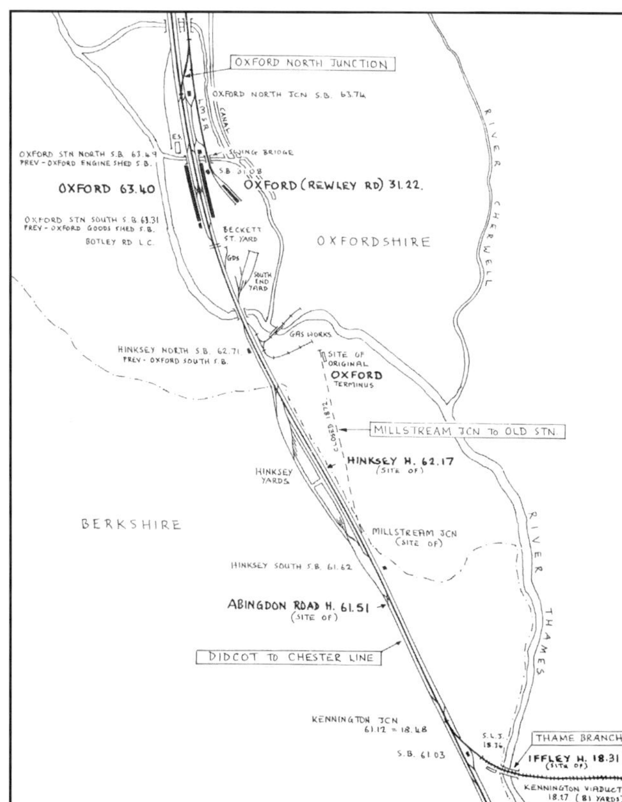

Right: An enlargement of the Oxford area taken from the *Atlas of the Great Western Railway as at 1947* by R A Cooke.

FROM THE ARCHIVES:
4-4-2 TANK - No. 4600

The designs attributed to G J Churchward have been extensively documented and generally speaking curiosities can be logically explained. As discussed in *Western Times* Issue 6, one example was the proposition that the effectiveness of 2-6-2T Class 45xx would have rendered unnecessary the creation of 2-6-2T Class 39xx but for a question of timing. In the ranks of modern mid-range tank locomotives however, there is one case that remains enigmatic without clear definition concerning its intended role.

No. 4600 was completed in November 1913 and is thought to have been intended as a possible successor to the ageing 2-4-0T 'Metro' Class on suburban services over routes where gradients were less arduous. This description in itself prescribes the confinements of its possible duties. Suburban train weights were increasing through growing custom and heavier non-corridor bogie coaching stock while tighter schedules added to the imperatives of good adhesive qualities and acceleration away from frequent station stops. Ten-wheeled four-coupled passenger tanks had been introduced in 1905 in the form of Class 2221 and were tried in Devon, the Wolverhampton district and some other areas where they were unsuccessful. With their 6' 8.5" driving wheels, the 'County' tanks were used mainly on Paddington-Reading express services plus semi-fast workings Reading-Oxford, Paddington-High Wycombe, and slower trains to Aylesbury and Hungerford.

Very likely from the outset, the 'Large Prairies' demonstrated the necessary competence to handle this work and 'County' tanks were replaced in 1931-5 largely due to introduction of 2-6-2T Class 61xx which was designed especially for suburban traffic radiating from the capital. The 61s remained masters of this work until succumbing to dieselisation.

WESTERN TIMES

A parallel situation confronted No. 4600 where the lively performance of the small prairies of Class 45xx delivered good starting capacity and an impressive turn of speed. It was hard to see how and under what circumstances, the little 4-4-2T could deliver a substantive improvement in performance. Dimensional comparison with Class 45xx:

Class	45xx	4600
Introduced	1906	1913
Wheel Arrangement	2-6-2T	4-4-2T
Running Numbers	4500-74	4600
Cylinders	17" x 24"	17" x 24"
Wheel Diameters		
- pony/ bogie	3' 2"	3' 2"
- driving	4' 7.5"	5' 8"
- trailing	3' 2"	3' 2"
Boiler Pressure (lb/ sq in)	180	200
Heating Surfaces (sq ft) §		
- firebox	96/ 94	94/ 94
- tubes	1178/ 1020	1178/ 1020
- superheater	na/ 102	na/ 102
Grate (sq ft)	16.8	16.6
Boiler Type	Std No. 5	Std No. 5
Tractive Effort (lb - 85%)	19,120	17,340
Weight (tons)		
- maximum axle loading	14.5	16
- adhesive	43.5	30.75
- total	57	60.6
Wheelbase	8' 4" + 5' 6" + 6' 0" +7' 0"	7' 0"+ 7' 0" + 7' 0" + 7' 0"
Water Capacity (gallons)	1000	1090
Factor of Adhesion §§	5.1	3.6

§ Differences relate to installation of superheating which was from Aug 1913 with Class 45xx and in Dec 1918 for No. 4600.

§§ Defined as the adhesive weight in lb divided by the tractive effort; the ideal median for British conditions was considered to be four. The lower the factor, the greater the propensity to slip.

No. 4600 was an attractive and well-proportioned locomotive, its looks enhanced by curved framing front and rear, and by provision of a copper-capped chimney. It was distinctive in being the only front bogie Churchward design to be fitted with front struts.

Adhesion problems indicated by the dimensions quoted above were confirmed in practice. The only partial remedy available to footplate crews was frequent pauses to take on water to maximise the weight borne by the side tanks. An attempt to improve weight distribution was affected in May 1915 whereby the bogie centre was moved forward by four inches which resulted in the following changed dimensions:

Weight (tons)	
- maximum axle loading	16.35
- adhesive	31.7
- total	60.35
Wheelbase	7' 0"+ 7' 4" + 7' 0" + 7' 0"
Factor of Adhesion	3.7

All these modifications failed to achieve any significant improvement in performance. The locomotive was originally deployed on suburban duties in the Birmingham area but moved to west Wales to work the Pembroke Dock branch in about 1918. It was still allocated to Neyland or Whitland when withdrawn in July 1925 with a mileage of 248,000. No. 4600 was camera shy and the accompanying photographs are the only examples readily to hand. A rare view in active service showed the engine on the traverser at Birmingham Moor Street. This was seen by the archivist some years ago but assiduous search of the files has proved unsuccessful. This elusiveness adds to the air of mystery surrounding this handsome machine; any views showing it in service would be most welcome for publication.

Above: This three-quarter view emphasises the elegance of the design; if only it had been of more practical use, on looks alone more examples of the type would have been welcome additions to the fleet.

Left: As running in 1924. Note that the garter coat of arms seen in the works portraits was never actually carried in service.

BRANCH LINE TO WALLINGFORD

The ancient town of Wallingford in Oxfordshire grew around a crossing point on the River Thames, situated roughly equidistant between Oxford and Reading. It boasts a distinguished history for a settlement of its size, having played a significant role in both the Norman Conquest and the English Civil War.

Its railway history began on 2 July 1866, with the opening of the Wallingford and Watlington Railway, the first stage of a proposed line connecting Brunel's Great Western Railway at Cholsey with the Wycombe line at Princes Risborough. This aspiration was to fail to proceed beyond Wallingford, with the original company selling out to the GWR in 1872. It was however, noted for being the first branch off the GWR main line built to the narrow (standard) gauge, and was 2 miles 52 chains in length.

Throughout its life, the line operated on a 'one engine in steam' basis, with the locomotive (supplied by Reading shed) out-stationed at Wallingford. Armstrong '517' Class 0-4-2Ts were the staple motive power, until succeeded by 48xx tanks from 1935. Passengers were always served with a regular service throughout the life of the branch, which was augmented with a prosperous goods trade, consisting of coal, livestock, milk, grain, sugar beet and various military supplies for RAF Benson.

The branch officially closed to passenger services on Monday 15 June 1959, although the last train was to run on the preceding Saturday with No. 1444 and a strengthened six-coach rake. Goods services to the yard at Wallingford were to soldier on until Monday 13 September 1965. Final closure occurred in May 1981, with the removal of traffic to the Associated British Maltsters (ABM) private siding.

This brief photographic homage presents a selection of images of the short branch from The Transport Treasury archives.

The compact station building at Wallingford is seen from the forecourt in 1958. The original red brick structure was augmented with an 1891 extension to the right-hand side, providing an enlarged booking office. The tower behind belonged to the town waterworks. ***James Harrold (H336).***

Above: During a trip along the branch on 14 May 1951, Dick Riley captured the full sequence of operational activity involving the auto-train, including the detached 0-4-2T No. 1447 being serviced on the shed road at Wallingford between turns. *R C Riley (RCR 3135).*

Below: A more unusual form of motive power is seen in the shape of an unidentified Collett '2251' Class 0-6-0. Supplied by Didcot shed, these tender engines began to appear following closure of the small shed at Wallingford in 1956, breaking the 0-4-2T domination.

Top: The rather neatly proportioned Wallingford signal box was a simple wooden cabin sat upon a red brick base, with a slate roof. The relatively simple track layout at the terminus site only required the provision of 13-levers within the box. Although an original signalling diagram dated 9 June 1892 still exists, it is not recorded exactly when the cabin was constructed.

It closed on 19 January 1964, although for much of its final years had only been manned on an as required basis by the goods foreman when shunting was taking place in the yard. Now long since swept away along with all the other buildings at Wallingford, one wonders if the 'Wallingford Station Box' plate survived, maybe a reader knows? *James Harrold (H338).*

Middle: Noted photographer Leslie Freeman was a regular visitor to the branch over a number of years, originally initiated by his period of National Service at nearby RAF Benson. During these visits he recorded many valuable scenes of the infrastructure and buildings, such as this view of the well stocked Goods Yard taken on 27 April 1957. The resident AA19 20T 'Toad' Brake Van No. W114898 was suitably branded 'Wallingford R.U. Not For Common Use'

Freight traffic was withdrawn from Wallingford on 13 September 1965, although trains did continue to run to the Associated British Maltsters site (adjacent to the current preserved terminus) south of the original station, until final closure in May 1981. *Leslie Freeman (LRF 2739).*

Bottom: During the penultimate summer of passenger operation, No. 1407 departs Wallingford with a late afternoon service on 6 September 1958. This particular Collett 0-4-2T was a branch stalwart, having arrived at Reading in late 1935 and remaining there until transfer to Southall in February 1960.

The cattle pens remain in-situ to the right, as does the 3000-gallon conical water tank which was fitted in 1920. Coaling took place from the rudimentary platform still seen to the left of the locomotive, or often directly from a loco coal wagon stabled on the Gas Works siding before its removal.

The whole of the original station site is now lost under a housing development, although the gabled villas and trees to the rear remain, on the still appropriately named Station Road.

A E Bennett (AEB 4177).

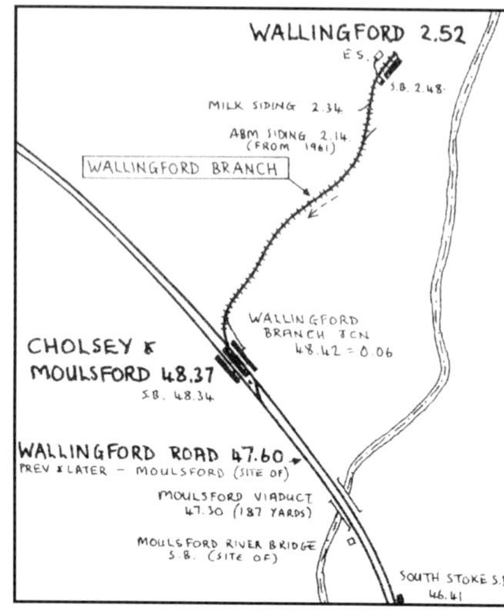

The Wallingford Branch (Map 31) *R A Cooke.*

Above: A rather forlorn looking scene of the station platform and former engine shed at Wallingford taken during the early 1960's. The basic brick built shed dated from around 1890 and closed on 11 February 1956, with the track removed shortly after. *Eric Miles.*

Right: On Sunday 9 April 1961, the London Branch of the Railway Correspondence & Travel Society (RCTS) ran the 'Berks & Wilts Railtour' from Paddington. The tour traversed the Malmesbury, Highworth and Farringdon branches, before visiting Wallingford. The six-car (2x3) Swindon Cross-Country Class 120 DMU is captured here at the terminus, prior to returning to Didcot. *Alec Swain (K2/5).*

Left: George Armstrong '517' Class 0-4-2T No. 1479 comes off the Wallingford branch and onto the Up relief line at Cholsey and Moulsford in January 1931. It is possibly engaged in the weekly locomotive change-over from Reading, which occurred on a Monday. Built at Wolverhampton in November 1884, the auto-fitted tank was allocated to Wallingford from 1930 until withdrawal in April 1932.

The branch line curved away to the right behind the locomotive and the sidings beyond the track gang (left centre) were utilised for the interchange of goods traffic generated at Wallingford. This traffic increased following the opening of the CWS Creamery siding in early 1934.
Dr Ian C Allen (GW 96).

Above: Branch regular 0-4-2T No. 1407 stands in the Wallingford bay platform at Cholsey in 1958, attached to Dia. A34 57ft Brake Third Autotrailer No. W1670W. The coach was part of a batch of four (Lot.1600), built at Swindon in April 1939. *James Harrold (H334).*

Below: Cholsey and Moulsford station is seen on 14 May 1951, with 'Hall' Class 4-6-0 No. 6937 *Conyngham Hall* slowing to call. The train is most likely an Oxford stopping service, with the connecting branch autotrain waiting in the bay. *R C Riley (RCR 3135).*

THE GREAT WESTERN TRUST (GWT) - BULLETIN NO. 8

Having an ever growing Collection means that items we acquired by generous donations in 1986, can be complimented by related ones acquired almost 40 years later. A common factor that such connections reveal however is that as has been said, 'the Past is a Different Country'.

The item I begin with was donated to the Trust in 1986 by the family of its author and is a wonderful example of just how untapped are family ephemera of a seemingly transient kind, but which, having been kept for many years, can now cast a unique light upon contemporary railway travel.

Illustrated is a scan of the official headed notepaper, folded to four sides no less, that the GWR provided in their Edwardian Restaurant Cars. It was used by Alfred Burne, then a subaltern in the Royal Horse Artillery when returning from fox hunting near Weymouth, still in his Saville Row riding gear, on the train from Westbury to Paddington on 29 December 1911. His reason was to provide his father with a commentary on the 3 shilling fee dinner he was taking (the superior menu at 3/6d had pheasant and a dessert) whilst adding anecdotes about his hunting that day. It is in hurried pencil. We have a complete transcript which is so of its time, his social status and his expectations, but I can only offer snippets here! He was being rushed by the steward as a second sitting awaited, but he firmly resisted! The fish course was stone cold! He remarked to the steward that the movable chairs were uncomfortable and coach riding not a patch upon the LNWR ones, to which the steward replied "So I believe". That said, this tiny comment proves it was in just one of a pair of GWR carriages (Nos. 405 & 406) that had such chairs, when the remainder had fixed bench seats. From a bad start, Burne then commented favourably on all remaining courses! He later served in WW1 as a Lieutenant-Colonel, receiving the very first DSO, and in later life (dying in 1959) he became deeply interested in military history. Contrary to his views on the carriage, the earlier *GWR Staff Magazine* of October 1906 states '...we understand that these new cars are giving unqualified satisfaction'. Umm.

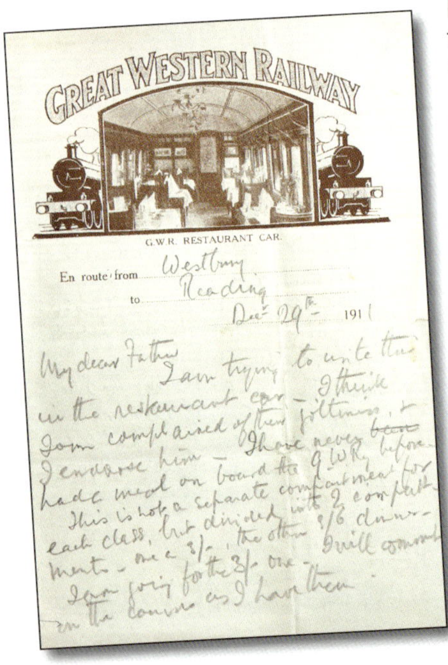
GREAT WESTERN RAILWAY
G.W.R. RESTAURANT CAR.
En route from Westbury
to Reading
Dec 29th 1911

Among the minutia of detail our Collection provides to volunteers like me who thrive on such things, I can add that these carriages had a peculiar design to our thinking, in that it had a curtain rather than swing door to separate the end kitchen from the dining area. So what? Well, a later revised *Restaurant Car Rule Book* we hold specifically states that it was essential to keep a 'door' closed between the kitchen and dining area, especially when cooking vegetables. Passenger complaints no doubt!

The final illustrations, similarly relate to GWR Dining Cars services, but in the form of an Official GWR Postcard, with a sepia vignette of Tenby on its face, but the reverse shows it's a Dining Car card. It was used, again in longhand pencil perhaps by an aunt to her niece 'Winnie' from the express to Cornwall at 2pm on 2 August 1908, having just passed Newton Abbot. It was posted in Plymouth. The pencilled annotation at the top, is the unique GWT Accession Identity Number for it, an example of our cataloguing regime to ensure all donation provenance is secured in our database.

This postcard is but one of seven in this 1906 series focussed upon GWR seaside resorts, which were followed in 1913-17 by a Restaurant Car & Correspondence Card series.

Both of these examples hark back to an unimaginable era of high class dining car travel with a correspondence service included! What chance of that today? Definitely, it was a very different country.

Peter Rance - GWT Trustee & Collection Manager.

THE GUARD'S COMPARTMENT

ISSUE 4

Ian Hargrave corrected the date of the 'Warship' **Front Cover** photo as being taken in August 1959, the loco having just removed the trains dining car portion, before proceeding west into Cornwall. He knows this for certain, as he is the young boy in the foreground!

Brian Williams contacted us to inform that in regards to the further discussion of the cover image from Issue 1 on **Page 78,** the driver on the 'Capitals United Express' on 27 June 1960 was Eddie *Broome*, not *Brown* as stated. As Chairman of the Canton Loco LDC, Eddie had sufficient clout with the roster clerk to ensure he was on the run that day. Incidentally, although a prestige working the job was not in the top link, merely in the vacuum goods link. After No. 92220's return to normal work, it worried the running foreman that it had to be rostered to an out and back turn especially if it was Midlands bound, in case the dastardly LMR got hold and kept it!

--- o O o---

CASTLE CLASS CENTENARY SPECIAL

Brian Williams also commented on the lower caption on **Page 59** of No. 7001 *Sir James Milne*. He correctly identifies it is not on its final run into Paddington as described, but is in fact still trailing coupled to an empty coaching stock move from Paddington. The loco is in back gear, the driver is reading a newspaper, and the fireman is seen between the tender and the last coach uncoupling. The train is standing on the down goods and carriage line, and as the photograph is consecutively numbered with the one above of No. 7000 *Viscount Portal*, they are likely to be of the identical train. What chance sequentially numbered 'Castles' coupled to either end of the same rake of coaches!

The image of No. 4079 *Pendennis Castle* at the foot of **Page 91** prompted **Mike Romans** to comment on the West London Carriage Sidings that can be glimpsed to the left of frame. 'Alas it seems to be a much overlooked location notwithstanding its considerable importance as a passenger yard, even including DMU servicing facilities. West London dealt with suburban stock, whilst Old Oak handled the coaches for mainline workings'.

--- o O o---

ISSUE 8

Ian Harrison made the following observations in relation to the photograph on **Page 29**, showing a 'Queen' Class 2-2-2 with a train of iron mink wagons.

The location is surely Witney Goods, the goods station being retained at the location of the former station, after the new passenger station was created when the branch was extended to Fairford. The train would be the 'Witney Blanket'. There are several pictures of a similar nature taken over the years. The trainload of blankets was taken from Witney, where they were manufactured, to Maples department store in London. A similar posed photograph from 1923, shows a train of 16 wooden Mink vans hauled by 3232 Class 2-4-0 No. 3247, while another shows such a train headed by one of the ex-Midland and South Western 2-4-0s.

So, I am sure that the purpose of these photographs was promotional - probably jointly by the GWR and Maples. Although they can't be made out in the picture on page 29, there were boards which indicated that the destination of the blankets was Maples. You can see that there is such a board to the left of the doors of the first wagon in the train.

Brian Williams noted that on **Page 50** the 'Hymek' wheel arrangement is B-B not Bo-Bo, both axles connected via secondary cardan shafts. This caused much head scratching when they were sent to LMR depots for tyre reprofiling, until they were told to jack up the rear of the bogie. Also the picture caption on the same page refers to the fitting of AWS, but should read GWR/BR(W) ATC.

On **Page 51** the 'Hymek' window surround colour is quoted as ivory-white, to be accurate the official Swindon Works 'Colour Chart' gives it as French Grey.

Above: 'Hymek' No. D7026 stands at Old Oak Common sporting French Grey window surrounds. Note also the snowploughs.

WESTERN TIMES

Just as **Issue 8** was due to be delivered from the printer, the writer had the privilege of meeting **Richard James**, son of the last Station Master at Marlborough. Richard is a true MSWJ enthusiast and when informed of the forthcoming article on **Page 74** kindly offered to send some additional information.

The first item sent was a reference to the *Railway Magazine* of 1899 which contained an article on the new Marlborough and Grafton Railway, which of course included the tunnel. The article dealt with the history of the railway and the associated MSWJ (the M&G had been sponsored as a separate company) and also included a number of illustrations of the M&G and MSWJ. One of these photographs is shown (left), the south side approach to the tunnel mouth. Whilst the image may be from a periodical of 125 years ago, and in consequence of reduced quality, it is still of particular interest as it shows the sheer sides of the approach cutting. Small wonder then that chalk falls were to occur later.

Richard also sent a letter giving further detail:

'My father, Fred James, took up the position of Station Master at Marlborough sometime in late 1958 following John Barnby's move to Witney. He told me that on one occasion whilst passing through the tunnel on his way to Savernake Low Level he was overtaken by a train heading in the same direction and it "frightened the life out of him". Alas, my curiosity with this story only grew after his death in 1993. I have no reason to doubt him and can only conclude he was on the 4pm shuttle to Savernake LL one Saturday afternoon when a late running Cheltenham to Southampton service due off Marlborough at 3.50pm rattled past on its way to Savernake High Level. The weak overline bridge (carrying the M&G over the GWR) at Wolfhall led to the closure of the High Level route in September 1958 and so maybe Dad had gone to Marlborough prior to that to look the place over.

'The most obvious question I never asked him was why on earth he took the job in the first place. Following the timetable 'cull' in June 1958 he must have had some foreboding that full closure was on the cards. John Barnby surely saw it coming!

'I travelled through the tunnel on many occasions in my youth, by rail, on foot and by car over the trackbed! Prior to closure in 1961 I would often travel over to Savernake, usually with the Guard, to spend a few hours in the company of the Station Master's son Christopher and knock off a few more numbers in my Ian Allan ABC as it was a great train-spotting location. Approaching the tunnel on the way home I would feel a certain apprehension staring up at that towering cliff of chalk. Following closure, it would be on our walking route to a field at the other end to pick mushrooms and no doubt as we were passing taking aim at the arms of the pair of up distant signals with the plentiful supply of ballast.

'My last association with the tunnel was between 1967 and 1969 when on winter Wednesday afternoons we would run across the top of it as part of the school's cross country course. This involved entering via the boundary fence on the up side, running across the top of the tunnel mouth and exiting on the down side. Nobody came to grief but I doubt the route would pass a 2024 risk assessment'!

Above: A number of years following closure of the line, this is the approach to the north end of the tunnel. Had the railway remained open serious money would have had to have been spent on stabilising and clearing the cutting sides. ***A E Bennett.***

--- o O o---